Chad A. Davis

Automated protein complex modelling

Chad A. Davis

Automated protein complex modelling

Computational assembly of macromolecules using heterogeneous structural templates

Südwestdeutscher Verlag für Hochschulschriften

Impressum/Imprint (nur für Deutschland/ only for Germany)
Bibliografische Information der Deutschen Nationalbibliothek: Die Deutsche Nationalbibliothek verzeichnet diese Publikation in der Deutschen Nationalbibliografie; detaillierte bibliografische Daten sind im Internet über http://dnb.d-nb.de abrufbar.

Alle in diesem Buch genannten Marken und Produktnamen unterliegen warenzeichen-, marken- oder patentrechtlichem Schutz bzw. sind Warenzeichen oder eingetragene Warenzeichen der jeweiligen Inhaber. Die Wiedergabe von Marken, Produktnamen, Gebrauchsnamen, Handelsnamen, Warenbezeichnungen u.s.w. in diesem Werk berechtigt auch ohne besondere Kennzeichnung nicht zu der Annahme, dass solche Namen im Sinne der Warenzeichen- und Markenschutzgesetzgebung als frei zu betrachten wären und daher von jedermann benutzt werden dürften.

Verlag: Südwestdeutscher Verlag für Hochschulschriften GmbH & Co. KG
Dudweiler Landstr. 99, 66123 Saarbrücken, Deutschland
Telefon +49 681 37 20 271-1, Telefax +49 681 37 20 271-0
Email: info@svh-verlag.de
Zugl.: University of Heidelberg, Dissertation, 2010

Herstellung in Deutschland:
Schaltungsdienst Lange o.H.G., Berlin
Books on Demand GmbH, Norderstedt
Reha GmbH, Saarbrücken
Amazon Distribution GmbH, Leipzig
ISBN: 978-3-8381-2376-9

Imprint (only for USA, GB)
Bibliographic information published by the Deutsche Nationalbibliothek: The Deutsche Nationalbibliothek lists this publication in the Deutsche Nationalbibliografie; detailed bibliographic data are available in the Internet at http://dnb.d-nb.de.

Any brand names and product names mentioned in this book are subject to trademark, brand or patent protection and are trademarks or registered trademarks of their respective holders. The use of brand names, product names, common names, trade names, product descriptions etc. even without a particular marking in this works is in no way to be construed to mean that such names may be regarded as unrestricted in respect of trademark and brand protection legislation and could thus be used by anyone.

Publisher: Südwestdeutscher Verlag für Hochschulschriften GmbH & Co. KG
Dudweiler Landstr. 99, 66123 Saarbrücken, Germany
Phone +49 681 37 20 271-1, Fax +49 681 37 20 271-0
Email: info@svh-verlag.de

Printed in the U.S.A.
Printed in the U.K. by (see last page)
ISBN: 978-3-8381-2376-9

Copyright © 2011 by the author and Südwestdeutscher Verlag für Hochschulschriften GmbH & Co. KG and licensors
All rights reserved. Saarbrücken 2011

Abstract

Protein interaction networks provide an increasingly complex picture of the relationships between macromolecules in the cell. Complementing these interactions with structural data provides critical insights into interaction mechanisms. However, structural information is available only for a tiny fraction of protein interactions and complexes currently known. To address this gap, we have developed a method to predict macromolecular complex structures by systematic combination of pairwise interactions of known structure. We first identify all interactions within a network that are of known structure or sufficiently similar to known structure to permit homology modelling. We then use these structural constraints to construct models of complexes. We tackle combinatorial explosion by developing an efficient algorithm that exploits heuristics to reduce the large search space and complement this with an automated scoring system to filter out the exponentially large number of unrealistic complexes, leaving a ranked set of the most plausible models. To test the approach, we defined a benchmark set of complexes of known structure, and show that many complexes can be re-created with good accuracy, using templates below 75% sequence identity. Certain models are much larger and more complete than what is capable with traditional modelling techniques. The approach can identify the most plausible homology models for a complex of dozens of proteins in less than a few hours. We applied the approach to whole-proteome sets of complexes from *S. cerevisiae*. For the complexes of known structure, we are able to identify the native complex in the majority of cases. We provide promising models for several dozen additional complexes, including multiple isoforms for each. Modelled complexes also provide functional classification, particularly for unannotated complexes from structural genomics initiatives. We show that the best results are achieved when the stoichiometry of the components is known and when the modelling is approached hierarchically, where core components, representing high-confidence interactions, are modelled before non-obligate interactions. We are refining this aspect of the automated modelling and making the procedure publicly available via a web service, to aid in the analysis of models. As the rate of structurally resolved interactions grows, our ability to model larger and more diverse complexes will grow exponentially.

Contents

1 Introduction..7
    1.1 Determining interactions...8
    1.2 Determining complex composition...8
    1.3 Determining macromolecular structure 10
    1.4 Modelling interfaces 11
    1.5 Modelling multimeric complexes..13
        1.5.1 Filtering exclusive interactions 14
        1.5.2 Electron microscopy density fitting...14
        1.5.3 Combinatorial docking 15
        1.5.4 Superposition of shared components...15
    1.6 Approach and applications...16
2 Methods..17
    2.1 Structured interaction database..17
    2.2 Structured interaction networks..18
        2.2.1 Searching pairs of sequences..19
        2.2.2 Verifying contacts...20
        2.2.3 Scoring interface templates..20
        2.2.4 Identifying redundant templates 21
    2.3 Interaction network traversal...21
        2.3.1 Measuring computational complexity...22
        2.3.2 Traversing an interaction network..23
        2.3.3 Merging complexes with shared components..23
        2.3.4 Identifying exclusive interactions..24
        2.3.5 Detecting collisions..25
        2.3.6 Detecting ring topologies...26
    2.4 Scoring modelled complexes..29
    2.5 Clustering redundant models..30
    2.6 Filtering steric clashes..30
3 Benchmarking modelled complexes..31
    3.1 Defining a non-trivial benchmark...31
    3.2 Comparing a model to a benchmark complex..35
    3.3 Avoiding parameter bias...37
4 Benchmark results..39
    4.1 RMSD threshold for correctness..39
    4.2 Modelling coverage..39

- 4.3 Accuracy ... 43
- 4.4 Ranking models ... 45
- 4.5 Sequence identity threshold of modellability ... 46
- 4.6 Weights of model characteristics ... 47
- 5 Defining the yeast complexome ... 49
  - 5.1 Structured interface templates ... 49
  - 5.2 Structure of individual components ... 52
- 6 Results of yeast complex modellings ... 53
  - 6.1 3D Repertoire coverage ... 53
  - 6.2 CYC2008 coverage ... 55
  - 6.3 Reconstruction of known complexes ... 56
  - 6.4 Predicted complex structures ... 60
- 7 Discussion ... 65
  - 7.1 Scoring ... 65
  - 7.2 Complex composition ... 66
  - 7.3 Stoichiometry ... 67
  - 7.4 Docking templates ... 67
  - 7.5 Clash detection ... 68
  - 7.6 Interaction conservation ... 69
  - 7.7 Potential functional insights ... 70
  - 7.8 Alternative conformations ... 71
  - 7.9 Nucleic acids ... 71
- 8 Conclusion ... 73
  - 8.1 Current work ... 74
    - 8.1.1 Web services ... 74
    - 8.1.2 Atomic modelling ... 75
    - 8.1.3 Refined clash detection ... 75
    - 8.1.4 Defining core complexes ... 76
    - 8.1.5 Hierarchical assembly ... 76
    - 8.1.6 Additional interaction data ... 77
    - 8.1.7 Novel interaction candidates ... 77
  - 8.2 Resources and tools developed ... 78
    - 8.2.1 Usage scenarios ... 78
    - 8.2.2 Software libraries ... 79
  - 8.3 Contributions ... 79
- 9 References ... 81

Figures

Fig 1.1: Interaction discovery methods..8
Fig 1.2: Clustering yeast TAP-MS interactions...9
Fig 1.3: Identifying the biological unit from X-ray structure..................................10
Fig 1.4: Applications of (monomeric) homology models at various levels of accuracy..............12
Fig 1.5: Transferability of functional annotation of protein-protein interactions 13
Fig 1.6: Docking within EM maps..14
Fig 1.7: Linking shared components of interactions..15
Fig 2.1: Template search procedure..19
Fig 2.2: Number of possible complex models...22
Fig 2.3: Traversing an interaction network..23
Fig 2.4: Merging complexes with shared components..24
Fig 2.5: Collision detection..25
Fig 2.6: Number of ring topologies in protein complexes......................................27
Fig 2.7: Ring detection..28
Fig 2.8: Scorable characteristics of complexes...29
Fig 3.1: Sizes of benchmark complexes..32
Fig 3.2: Classifications of benchmark complexes...33
Fig 3.3: Genera of benchmark complexes...34
Fig 3.4: Misalignment of complexes resulting from internal homology.................35
Fig 4.1: Interpreting RMSD between complexes...40
Fig 4.2: Coverage of benchmark set of complexes...41
Fig 4.3: Size of sub-complex models..41
Fig 4.4: Best-scoring complete models...42
Fig 4.5: Correlation between backbone RMSD and prediction score...................43
Fig 4.6: ROC curve..44
Fig 4.7: Contingency matrix...45
Fig 4.8: Rank of the best-scoring model per benchmark target............................45
Fig 4.9: Sequence identity limits of modelling..46
Fig 4.10: Score weighting determined by ordinary least squares (OLS) 47
Fig 5.1: Socio-affinities between complex components.......................................49
Fig 5.2: Structured interaction network of S. cerevisiae.......................................50
Fig 5.3: Sizes of 3D Repertoire complexes..51
Fig 5.4: Sizes of CYC2008 complexes...51
Fig 5.5:Sources of yeast interface templates...51
Fig 5.6: Using known structures and homology models......................................52
Fig 6.1: Modelling coverage of 3D Repertoire complexes...................................53

Fig 6.2: Modelled complexes: 3D Repertoire...54
Fig 6.3: Modelling coverage of CYC2008 complexes..55
Fig 6.4: Modelled complexes: CYC2008...56
Fig 6.5: Proteasome models...57
Fig 6.6: Cytochrome-bc1 (Complex III)...58
Fig 6.7: COPII model..61
Fig 6.8: Arp2/3 model..61
Fig 6.9: cAMP dependent protein kinase 62
Fig 6.10: Methionyl glutamyl tRNA sythetase..62
Fig 6.11: TRAPP complex extended via docking..63
Fig 7.1: Gloebacter violaceus (GLIC) ion channel..69
Fig 7.2: Templates for unannotated benchmark target...70
Fig 7.3: Chlorite dismutase-like family 71
Fig 8.1: Web application..74
Fig 8.2: Structural bioinformatcs libraries...79

1 Introduction

The function of a protein in the cell is largely defined by how it interacts with other proteins (Yu et al. 2004). Whole-genome studies have provided inventories of many of the interactions in an organism (Krogan et al. 2006; Gavin et al. 2006) and interaction data from diverse experiments continues to increase. Networks of these interactions have identified which protein hubs are essential for one or more pathways (Ning et al. 2010). Despite this flood of new data, our level of mechanistic understanding of these interactions is not keeping pace. That specific proteins interact does not yet tell us how they interact at the molecular level. Using protein structures to complement networks can provide more insight into how these interactions function at the molecular level (Aloy & Russell 2002). Structural data can also provide constraints for determining which interactions do not take place simultaneously (Jung et al. 2010) and therefore sharpen the distinction between obligate interactions in stable complexes and those proteins that participate in multiple pathways (Kim et al. 2006; Gavin et al. 2006). This knowledge is relevant to annotating pathways and to studying drugs that target protein-protein interactions (Brooijmans et al. 2002).

Structural genomics initiatives and the increased pace of experimental structure determination have provided detailed atomic structures for many of these proteins, interactions and complexes (Chandonia & Brenner 2006). These experimental data, while invaluable, are only individual pieces of a complex system. Ultimately, it is the synergistic effect of a set of simultaneously interacting proteins in a complex that defines the functional modules driving biological pathways.

Our goal in this project is to integrate structures of interactions into networks; this involves not only isolated interactions but the determination of which interactions take place simultaneously within macromolecular complexes. Since proteins interact with multiple partners, the complexome of a species is an order of magnitude larger than the interactome (Karaca et al. 2010). Furthermore, a complex is not static but exists in several different forms, with different interaction partners of varying duration. We aim, ultimately, to provide atomic models for the structures of all of these complex forms. We accomplish this by identifying sets of structurally mutually compatible interactions. The first requirement is to identify interacting pairs of proteins

1.1 Determining interactions

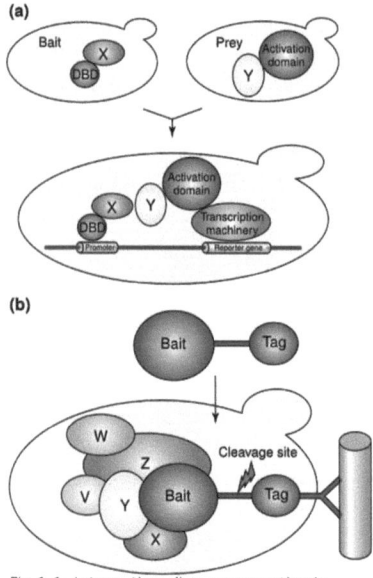

Fig 1.1: Interaction discovery methods
(a) The two-hybrid system and (b) affinity purification. DBD: DNA-binding domain. From (Aloy and Russell 2002)

There are many methods for determining functional (indirect) and physical (direct) associations between proteins (De Las Rivas & de Luis 2004; Shoemaker & Panchenko 2007a). Two of the most popular are the yeast two-hybrid (Y2H) system and tandem affinity purification (TAP) coupled to mass spectrometry (MS), also known as TAP-MS (Fig 1.1: Interaction discovery methods). Both Y2H (Ito et al. 2001; Uetz et al. 2000) as well as TAP-MS (Gavin et al. 2006; Krogan et al. 2006; Gavin et al. 2002; Ho et al. 2002) have been used in large-scale studies of the *S. cerevisiae* interactome. Increasingly, both techniques are applied to many other organisms (Stelzl et al. 2005; Rual et al. 2005; Kuehner et al. 2009).

Y2H has the advantage that it identifies direct, physical interactions, but it suffers from a high false positive and negative rates (Huang et al. 2007). In order to deduce direct interactions under TAP-MS, multiple rounds can be performed, in which in each protein in turn is tagged and used as the bait. This permits statistics on the strength of the association between two given proteins (Gavin et al. 2006). Recent advances in TAP-MS using chemical cross-linking, to keep subunits connected, have also permitted the identification of direct, physical interactions (Taverner et al. 2008). These interactions are recorded in a number of publicly accessible interaction databases (Rohl et al. 2006). The most comprehensive databases include: MINT (Ceol et al. 2010), DIP (Salwinski et al. 2004), IntAct (Aranda et al. 2010), and STRING (Jensen et al. 2008). Any set of interactions, whether from a single study or from a large interaction database implies an interaction network. A network, however, only collects the individual connections.

1.2 Determining complex composition

In order to identify functions within interaction networks, subsets of tightly interacting proteins can be clustered into complexes. There is wide range of clustering methods (Brohee & van Helden 2006) and different methods have derived different sets of complexes for the yeast interaction studies mentioned above (Gagneur et al. 2006). Many have attempted to regroup these into more consistent, tightly defined clusters (Collins et al. 2007; Pu et al. 2007; Ozawa et al. 2010; Krumsiek et al. 2008).

These approaches also differ in how they define a "cluster". Traditional clustering puts each protein into exactly one protein complex. This does not account for temporal aspects of interactions such as "moonlighting" (Jeffery 1999), whereby a single protein may play a role in multiple complexes. This has been addressed by making a distinction between core complexes, which are stable, and attachment modules, which are satellite sub-complexes, that may shuttle between core complexes, resulting in different complex isoforms (Fig 1.2: Clustering yeast TAP-MS interactions.) (Gavin et al. 2006). The interactions within a functional module are generally obligate and, therefore, occur simultaneously. In contrast, one functional module may have have many different non-obligate interactions with different partners that generally cannot occur simultaneously (Han et al. 2004). This is not a black-and-white distinction, however. There is a continuum between these extremes (Agarwal et al. 2010) and identifying exclusive interactions requires more than interaction networks alone. Knowing the structure of the components can show whether two interaction partners are able to bind simultaneously (Jung et al. 2010). Some clustering approaches assume a high-clustering coefficient, which is problematic with protein complexes where it is not necessarily the case that every protein in a complex contacts every other (Brohee & van Helden 2006). Due to some of the disadvantages of automated network clustering, others have focused on cataloguing complexes by manual curation of many small-scale interaction studies (Pu et al. 2009; Mewes et al. 2004).

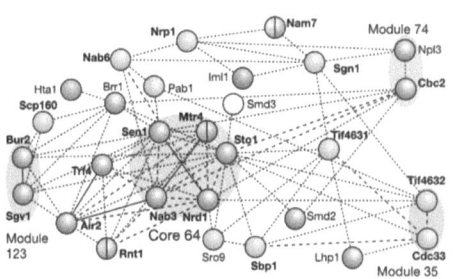

Fig 1.2: Clustering yeast TAP-MS interactions. "Core" denotes an obligate complex. A "Module" may interact with more than one core. From Gavin et al. 2006.

Knowing the protein components participating in a complex is still not the whole story. Protein complex evolution is strongly influence by subunit duplication (Levy et al. 2008; Taverner et al. 2008). Most methods for identifying interactions do not identify the copy number of each component in a complex. In the case of TAP-MS, one can identify whether a tagged protein also pulls down an untagged instance of the same protein (Kuehner et al. 2009). This suggests a homotypic interaction, but only identifies it as a plurality and does not determine the exact number of the components. Recent advances with chemical cross-linking and mass spectrometry of intact sub-complexes have started to identify the stoichiometry of complexes (as well as evidence for direct physical interactions) (Taverner et al. 2008), but not yet on a large scale. Nevertheless, direct contacts identify which proteins interact, but still do not specify the atomic details of how they interact. For that, it is necessary to determine the structure of the interactions.

1.3 Determining macromolecular structure

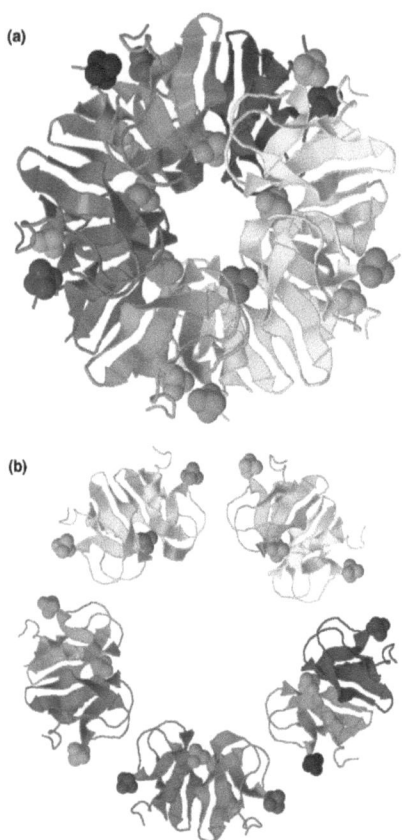

Fig 1.3: Identifying the biological unit from X-ray structure

a) Homodecameric trypsin inhibitor (PDB ID 2HEX) discovered to be five homodimers after eliminating crystal contacts. From Krissinel & Henrick 2007.

The most popular method for determining the structures for single proteins, interactions, and complexes is X-ray crystallography. In many cases, the resolution can reach the level of individual atoms. The method is limited by the requirement to crystallise protein samples, which is increasingly difficult for large, non-globular proteins or macromolecular complexes. However, techniques and automation are improving in terms of the quality and size of structures that can be resolved (Chandonia & Brenner 2006; Mueller et al. 2007). X-ray structures provide enough detail to be able to determine similarities between related domains (Murzin et al. 1995; Orengo et al. 2002). Interaction details down to the level of individual chemical bonds are possible. Structures of complexes provide the location of individual components and their relative orientations to one another. In principle, the structure of a complex also provides the stoichiometry of the components, assuming the quaternary structure, i.e. the biological assembly of the structure, is correct. This requires being able to differentiate between the contacts that are only an artefact of crystallisation and those which exist in the cell (Fig 1.3: Identifying the biological unit from X-ray structure). There are many methods for eliminating crystal contacts and deducing quaternary structure (Xu et al. 2006; Bordner & Gorin 2008; Henrick & Thornton

1998), the most recent is the PISA method (Krissinel & Henrick 2007). Currently, all determined X-ray structures are deposited into the Protein Databank (PDB) (Berman et al. 2007), which contains the Biounit set of biological assemblies, which is derived from manual and automatic annotation using PISA.

Another source of high-resolution structural data is nuclear magnetic resonance (NMR) spectroscopy. This method has the advantage that proteins are resolved in solution, rather than in a crystalline state. This makes NMR more amenable to studying dynamic processes, such as conformational changes. This is of particular interest as many such conformational changes are the result of protein-protein interactions. It is limited, generally, in the size of the structures that it can resolve, making it less capable of resolving multimeric complexes than X-ray crystallography, though combining the two methods can make this more tractable (Simon et al. 2010).

Electron microscopy (EM) and cryo-electron tomography (ET) are techniques that cannot currently reach the resolutions of X-ray crystallography, but can provide structures for much larger complexes or even entire cells (Baumeister 2005; Cyrklaff et al. 2007). While able to image multimeric assemblies several nanometres in diameter, the resolution generally does not reach that required for identifying contacts at interfaces. In the context of complex modelling, EM is most commonly used together with other experimentally determined structures (e.g. from X-ray or NMR), combined using various methods, to provide restraints on the overall quaternary structure of modelled assemblies. EM structures are often deposited into a public database, either the PDB or the Electron Microscopy Databank (EMDB) (Henrick et al. 2003), or both.

Nonetheless, there are still many complexes for which structures are currently unavailable. Structural genomics has contributed more single protein structures than multimeric structures. These are not only more challenging experimentally, but the size of the complexome of an organism is exponentially larger than its proteome (Karaca et al. 2010). Even when we have all the pieces, it will be much longer before we resolve all the states of all the puzzles.

1.4 Modelling interfaces

While there are many methods for predicting protein-protein interactions (Aloy & Russell 2006; Shoemaker & Panchenko 2007b; Valencia & Pazos 2002), the fact that two proteins interact only tells us their relative proximity in the complex. In order to be able to model the complex, we require structured interfaces for each interaction. As individual proteins belong to families (Chothia 1992), which provide the basis for homology modelling, so have interactions been shown to belong to discrete types (Aloy & Russell 2004), which provide the basis for homology modelling of interactions (Aloy et al. 2003; Teichmann 2002; Launay & Simonson 2008; Xu et al. 2006; Wong et al. 2008; Aloy & Russell 2002; Fukuhara & Kawabata 2008; Lu et al. 2003; Kundrotas & Alexov 2006). Transfer of functional annotation of interactions between species has been called *interologue mapping* (Yu et al. 2004), because it attributes functional annotation from an *interologue*, i.e. an interaction homologue, or homologous interaction (Walhout et al. 2000).

It has been estimated that interologues can provide structured interaction templates for 20% of known protein-protein interactions (Sinha et al. 2010). The exact coverage and quality depends on the conservation of each of the interacting proteins (Mika & Rost 2006; Saeed & Deane 2007). Some have cautioned against drawing structural conclusion from orthologous interactions, however (Park et al. 2004). As with homology modelling of monomeric proteins, the accuracy, and therefore the usefulness, of models decreases with decreasing sequence identity of the template structure (Fig 1.4: Applications of (monomeric) homology models at various levels of accuracy) (Baker & Sali 2001). Interactions are generally conserved between species at sequence identities above 30-40% (Yu et al. 2004; Aloy et al. 2003), except for mutations in specificity-determining residues. Below this threshold, the reliability of an annotation transfer drops quickly (Fig 1.5: Transferability of functional annotation of protein-protein interactions).

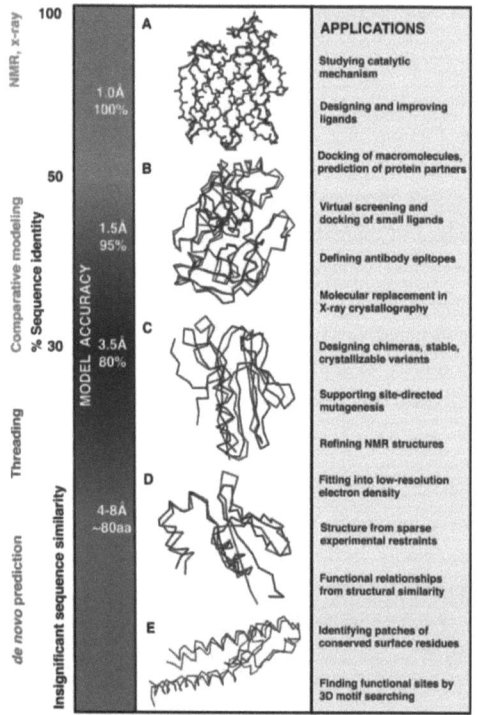

Fig 1.4: Applications of (monomeric) homology models at various levels of accuracy
From (Baker & Sali 2001).

For any pair of sequences there will generally be multiple potential interface templates. Some have showed that known structures contain enough alternative conformations to be able to model dynamic behaviour (Kohn et al. 2010). It is also important to realise that interacting proteins may change conformation upon entering their bound state, which is particularly challenging when predicting interactions with protein-protein docking (Zacharias 2010; Betts & Sternberg 1999).

There are many approaches to predict interaction interfaces using protein-protein docking (Lensink & Wodak 2010). Now there are also approaches that exist solely to combine an ensemble of docking poses into a smaller set of consensus predictions (Plewczynski et al. 2010). The docking field has recently seen a trend toward more knowledge-based approaches, where the pure physical parameters of docking methods are complemented by identifying homologous interactions in order to use these as a starting point in the search for the best modelled interface (Sinha et al. 2010; Karaca et al. 2010; de Vries et al. 2010; Guenther et al. 2007). It is not straight-forward to compare docking predictions to those from homology models, as they are based on different assumptions. In the docking case, the native unbound structures of the interactors are known, and their interaction orientation is predicted. In the homology modelling case, the interaction orientation is known and what is predicted is the extent to which the template structure represents the sequence. Despite this, the field will continue its general trend toward increasingly hybrid approaches to the structure modelling problem (Cowieson et al. 2008; Aloy et al. 2005; Alber et al. 2008).

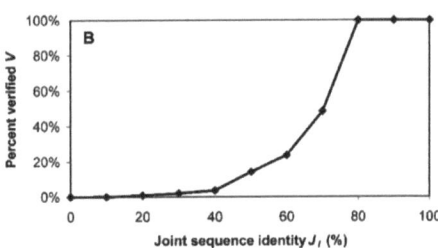

Fig 1.5: Transferability of functional annotation of protein-protein interactions
From Yu et al. 2004.

1.5 Modelling multimeric complexes

The next step beyond modelling dimeric protein interactions is to model multimeric complexes. However, this is not simply a repeated application of interface modelling, as there is a combinatorial increase, not only in which interactions are possible (N proteins may have up to $E = N \cdot (N-1)/2$ potential interactions), but in the number of interactions that may occur simultaneously (the number subsets of the set of interactions, $2^{|E|}$). This is complicated even further if one considers the level of internal symmetry of complexes and the preference for interaction types to be repeated within a complex (Levy et al. 2008).

1.5.1 Filtering exclusive interactions

The simplest approximate solution to this problem is to remove the interactions from an interaction network that are mutually exclusive (Kim et al. 2006). This can be done by identifying the boundaries of annotated domains on the interacting proteins and ensuring that these do not overlap in the sequence. This is an indirect way to prevent steric clashes, but can provide an initial filter against exclusive interactions (Ozawa et al. 2010). The next level is to consider the structure of the individual proteins and to locate the binding site and reject any binding sites that overlap as evidence for exclusive interactions (Jung et al. 2010).

However, a list of pairs of exclusive interactions does not provide a model for a complex. Again, all such pairs of interactions would have to be considered in order to identify a subset of interactions that are structurally internally consistent. Furthermore, steric hindrance does not occur only at an already-occupied binding site. A complex is more than a linear chain of proteins. An interface that induces a particular orientation of a large component protein may occlude binding sites from neighbouring proteins as well. This requires a view of the bigger picture.

1.5.2 Electron microscopy density fitting

Fitting high-resolution X-ray structures into low-resolution EM density maps is a well established method for localising protein components within complex volumes (Ceulemans & Russell 2004; Lasker et al. 2009; Lindert et al. 2009; Trabuco et al. 2008), even when the components themselves may be homology models (Topf et al. 2005; Topf et al. 2008). It has

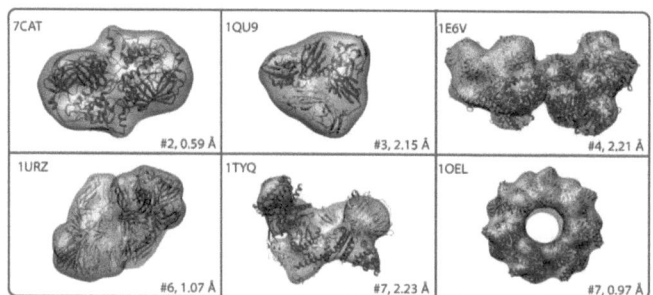

Fig 1.6: Docking within EM maps
#N identifies a complex with N protein components. From Lasker et al. 2010.

been used to provide models with atomic resolution for complexes that are too challenging for X-ray crystallography alone, such as the 26S proteasome (Förster et al. 2010) or the nuclear pore complex (Alber et al. 2007). EM fitting lends itself well to being combined with other methods. TAP-MS has been used to define relative proximities of components, which can then be fit into EM density maps (Alber et al. 2005). One of the most current approaches is to use EM maps as a volume constraint for combinatorial docking (Fig 1.6: Docking within EM maps) (Lasker et al. 2010).

1.5.3 Combinatorial docking

The first automated approach to combinatorial complex modelling was based on docking (Inbar et al. 2003; Inbar et al. 2005b; Inbar et al. 2005a). The emphasis is primarily on modelling symmetric complexes, with various types of symmetry. Heterotypic complexes are more challenging, requiring each interaction to be modelled specifically, whereas symmetric complexes are able to exploit repetitive interfaces that only need to be modelled explicitly one time. Complexes containing up to six and seven components can be modelled with current techniques (Karaca et al. 2010). These limits are the result not only of the computational cost of combinatorial assembly, but also the cost of finding the optimal docking itself. However, this step is not necessary when homologous interfaces exist.

1.5.4 Superposition of shared components

Homologous interfaces can also be combined combinatorially (Aloy et al. 2004; Aloy et al. 2005; Pichaud 2008; Taverner et al. 2008). This requires that two interfaces have a shared component, i.e. that one protein is shown interacting with two different interaction partners. Structural alignment of the common component places all three proteins into a single frame of reference. If no steric clashes result, this provides evidence for the two interactions taking place simultaneously and also results in a multimeric structural template (Fig 1.7: Linking shared components of interactions). There might be multiple such pairs of interactions containing shared components in an interaction network. This defines the

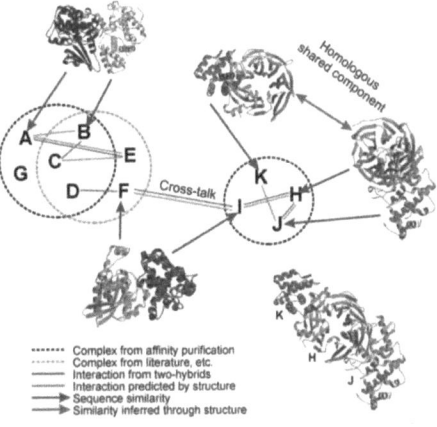

Fig 1.7: Linking shared components of interactions
From Aloy et al. 2004

combinatorial aspect of the problem. Additionally, however, there may be more than one interface template for a single interaction, i.e. the edges in the interaction network may actually be multi-edges, again multiplying the complexity. This is still distinct from the docking case, however, as each such interface template is derived from a native biological interface. The number of possible dockings is also orders of magnitude larger than the number of possible interface templates for a given pair of proteins. The final challenge is to rank the multitude of complexes modelled in a way that corresponds to their biological significance. This is an open problem, which we also address in our approach.

1.6 Approach and applications

In order to pass the current size limits of combinatorial assembly, we use a number of heuristics, including a novel algorithm for efficiently traversing an interaction network with multi-edges. Traditional graph algorithms that assume that edges are independent cannot be applied to spatial problems such as protein complex modelling (Jung et al. 2010). Structural consistency must be verified.

Our algorithm not only accommodates this, but also turns this into an advantage that helps us avoid billions of unnecessary verifications, significantly speeding up the assembly process. Any promising complex model can be used as a seed to iteratively add on additional interactions in subsequent modelling rounds, allowing ever larger complexes, and again saving significant search time. We are also able to build ring topologies, which no previous approach has done explicitly, and to verify when a ring closure is supported by structural evidence. This also permits us to identify candidates for potentially novel interfaces that arise from the assembly.

We demonstrate, by application first to a rigorous benchmark,and then to the yeast proteome, that the approach can often predict complex structures accurately, and identify potentially new structures in advance of structure determination by experimental methods.

2 Methods

Our approach to homology modelling of complexes is based on two high-level observations. The first is that structural models of interfaces can be built from interologues (Section 1.4 Modelling interfaces). The second is that interfaces, modelled or otherwise, can be combined into higher order structures via structural superposition of shared components. Such superimposed complexes can be chained incrementally, producing ever larger complex models (Section 1.5 Modelling multimeric complexes).

Modelling a complex can be broken down into three major steps. First, the composition of the complex, in terms of the components to be modelled, must be determined. Second, all applicable interaction templates for every potential interaction within a complex must be found. Third, pseudo-atomic structural models are assembled as the interaction network is traversed and sub-complexes are merged into larger complexes. This all depends on a searchable, curated dataset of structured interaction templates.

2.1 Structured interaction database

While there are many interaction databases (De Las Rivas & de Luis 2004), there are very few interface databases that provide atomic detail of the structure of an interaction. Without structure, even verified interactions cannot be used as interface templates for homology modelling.

Unlike the majority of interaction data sets, our set was not based on curated domain definitions such as SCOP (Murzin et al. 1995), CATH (Greene et al. 2007; Orengo et al. 1997), 3DID (Stein et al. 2005; Stein et al. 2008), or any of the many other domain databases. Rather, we seek to find interactions between structures that may be unannotated or where the interaction is facilitated by a segment smaller than an annotated domain. This is effectively a domain search in itself.

Using all of the structures in the Protein Databank (PDB) (Berman et al. 2007) (54858 structures, at the beginning of 2009). Sequences were extracted from each protein chain of each structure (136,710 unique sequences). We first searched for all conserved segments from known structures. This began with an all-against-all BLAST search (Altschul et al. 1997) of all the chains of all the complexes. High-scoring pairs (HSPs) that overlapped on a protein chain were grouped into blocks of at least 50 residues (to eliminate insignificant short overlaps). Blocks were grouped by single-linkage clustering when the BLAST E-value between them was ≤ 0.01. This defines a conserved segment. Sequence coordinates of segments were mapped to structural coordinates (i.e. to residue identifiers in the PDB structure coordinates) by aligning the native sequences to the sequences of the resolved structures. Proteins are frequently truncated to facilitate crystallisation and many other proteins have disordered residues or loop regions that are not represented in structures. Defining a mapping between residues in the native sequence and the corresponding residues in a structure allows moving back and forth between the two representations. Then, all of the identified structural segments of a cluster were structurally aligned using the superimposition method STAMP (Russell & Barton 1992).

Contacts between members of the segment families were identified using InterPreTS (Aloy & Russell 2003). Noting that interaction types are also repeated in nature (Aloy & Russell 2004), similar interfaces were clustered, with the distance metric being the interaction RMSD (iRMSD) with a threshold of 5Å, which was found to be the threshold for interaction similarity (Aloy et al. 2003). This resulted in 926642 Interactions, of which 247017 can be clustered by single-linkage with an iRMSD less than 5Å into 11069 groups, leaving 690694 searchable interface templates. The interactions here are between the conserved segments identified above, not between full-length proteins. Interactions are also counted once in each direction (e.g. *A* interacts with *B*, and *B* interacts with *A*; this makes finding and clustering interactions more efficient). This is why the number is larger than one might expect simply for clustered, full-length protein-protein interactions. This is also a consequence of the strict clustering at an iRMSD of 5Å. This is intended to remove only the most redundant interfaces. The construction of this database is largely automated so that it can be updated as new structures, particularly multimeric structures, are published.

2.2 Structured interaction networks

Given a set of protein components hypothesised to form a stable complex, the first step is to identify all interface templates that are homologous to any pair of proteins in the complex. We do not limit ourselves at this step to interactions with experimental evidence (e.g. yeast two-hybrid or tandem affinity purification data). The existence of a homologous interface for a pair of proteins is itself a strong indicator of a potential interaction. For each potential pair of interacting proteins, we identify multiple interface templates, in order to test multiple interface orientations.

Finding all templates within a complex requires doing all-against-all interaction modelling, i.e. for every potential pair of components in a complex. For N components, this is $N \cdot (N-1)/2$ searches for interaction templates.

2.2.1 Searching pairs of sequences

For each pair of sequences, a paired BLAST search of the PDB is performed. The BLAST database was created by extracting the C-alpha sequences of every chain of every structure from the PDB. This excludes unstructured regions and guarantees that any aligned positions will correspond to structured residues. Sequences are grouped at 100% identity. The BLAST E-value threshold was set to 0.01. The PSI-BLAST variant of BLAST was used with 2 iterations, keeping 500 hits per iteration. We do not generally set a sequence identity cut-off to prevent prematurely discarding distant similarities that may make more sense in the wider context of a complex model.

Since only proteins from the same structure can provide an interface template, the hits of each protein of a pair are then grouped to find pairs of hits occurring in a single PDB structure (Fig 2.1: Template search procedure). This simply eliminates any cases where a given structure is hit by only one sequence of the pair. This leaves a list of structures with at least one hit from each sequence in the pair. There may be more than one hit per query sequence, however. In this case the Cartesian product of the hits in a structure is formed, e.g. if a structure has two hits for the first sequence and three hits for the second, there are then six potential interactions to be checked in that structure. This gives all potential interaction

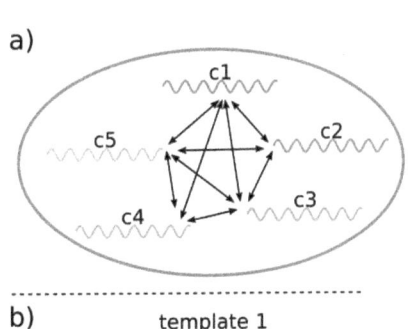

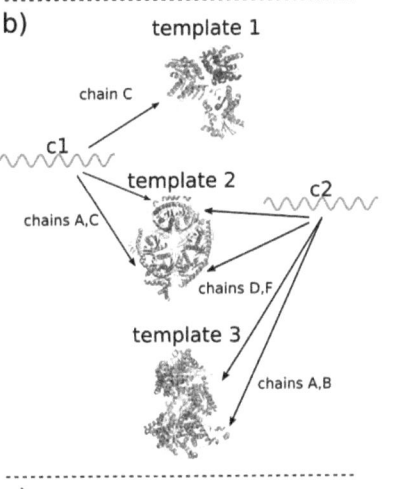

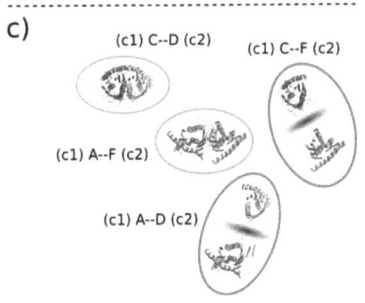

Fig 2.1: Template search procedure
a) all pairs of sequences to be modelled. b) template complexes identified by BLAST, hits on certain chains. If c1 hits A and C, and c2 hits D and F of template 2, then A--D, A--F, C--D, and C--F are all candidates. c) Contact check verifies only C--D and A--F as valid interface templates.

templates from a single structure for the given pair of sequences. The same is done for any other structures that could be found via sequence search for this pair. The same is done for every other pair of components in the complex to be modelled.

2.2.2 Verifying contacts

The sequence search finds structures containing segments homologous to our two query proteins (for every possible pair in the set of components). We then verify that these two structural templates are actually in contact. This is done using the database described above (Section 2.1 Structured interaction database). We first map the BLAST hits to the pre-computed segment families. Cases where both query sequences hit the same fragment of a structure are skipped, as these cannot represent dimeric interfaces. A minimum sequence coverage of 50% was required, in both directions. That is, the query sequence must align to at least 50% of the subject structure's sequence and vice versa. It is crucial in complex modelling, including interaction modelling, that a query sequence is not represented by a larger structural domain. This would lead to over-predicting the size of a structure. It is also suboptimal to use structural templates that do not cover a minimal percentage of the sequence, as this will mean that additional density is missing from the model which can be detrimental to the plausibility of the model, especially if this occurs at an interface.

This mapping is also one-to-many as there are multiple structured segments in each family. We again do the Cartesian product to produce all possible pairs of structured segments, which are potential interfaces if they are in contact. We then identify which pairs of segments are actually in contact, removing the others. This leaves us with a redundant set of interface templates. We then cluster the interfaces, using the pre-defined clusters (Section 2.1 Structured interaction database).

2.2.3 Scoring interface templates

The applicability of an interface template to model a pair of potentially interacting proteins is a function of the level of sequence conservation as well as the number of residues involved in the interface. As an interface template is effectively two templates, one for each of the two interacting proteins, these values exist twice for each interface template. We measure the average conservation ($cons_{avg}$) as well as the average number of residues in contact ($nres_{avg}$) and create a weighted average of those two values via:

$$weight = \frac{nres_{weight} \cdot nres_{avg}/10 + cons_{weight} \cdot cons_{avg}}{nres_{weight} + cons_{weight}}$$

where $nres_{weight}=0.1$ and $cons_{weight}=0.9$. In order to put conservation and interface size on the same scale, we assume that the maximum number of residues interacting on either side of an interface to be less than 1000 and scale this down by a factor of 10 to the range [0:100] (since conservation is measured as a percentage). As complex stability is a function of more than the interface size (Brooijmans et al. 2002), and because we do not want to be limited only to large interfaces, we attribute the majority of the interface template score to the sequence conservation. This score thus approximates the average level of conservation at an interface.

2.2.4 Identifying redundant templates

For each potential interaction, interface templates are clustered using the interface families defined above (Section 2.1 Structured interaction database). Generally, the highest scoring template in a cluster will be chosen to represent the cluster. This can be influenced by higher-level considerations, i.e. those that affect a complex as a whole, for example, preference for templates from a single source structure, preference for templates that have already been used elsewhere in the same structure, or preference for templates from a certain species.

Even after clustering interface templates, it is not possible to consider every interface family for every potential interaction in a complex. This is due partially to our strict definition of interface families. With smaller complexes, one can afford to dig deeper; with larger complexes one must settle for the top scoring interface templates. We compromise by considering the top 20 templates, proving 20 alternative interaction types, for each potentially interacting pair of proteins in a complex.

2.3 Interaction network traversal

The set of interface templates found above cannot simply be combined into a modelled complex. There are generally several interface templates for a given interaction, most of which are not correct. Interface templates for different interactions also may not be structurally compatible. Interactions cannot be considered independently, since one interface may prevent a different interaction from taking place simultaneously. This is the reason that classical graph algorithms, such as the minimum spanning tree, are not applicable to this problem: they assume an independence of the edges in the network. We require a method for selecting interfaces that will lead to a structurally coherent complex model.

2.3.1 Measuring computational complexity

It is not generally feasible to consider all possible complex topologies as this grows exponentially with the number of templates (Fig 2.2: Number of possible complex models). Thus we use the scores of the interfaces to rank the interface templates, which allows the most promising templates to be considered first.

As an interaction network may include many peripheral components (Gavin et al. 2006), it may not always be possible to include every component in every complex. We therefore also allow models for sub-complexes. The number of possible sub-networks defines the number of possible sub-complexes that can be modelled from a structured interaction network (Ozawa et al. 2010). The number of such sub-networks is the number of subsets of the edges in the interaction network. This is the power set of the set of edges, which contains 2^E subsets. In our case, it is important to point out that these edges are multi-edges, i.e. that each potential interaction in the network is represented by up to 20 structured interface templates. Each such interface template may have a different orientation, leading to a different complex model. If each edge has 20 such templates, then the number of modellable sub-complexes is, at most, $2^{20 \cdot E}$ (Fig 2.2 Number of possible complex models).

The complexity is slightly reduced by the fact that generally only one interaction will exist between a specific pair of proteins. It is also reduced by the fact that for many pairs of proteins, there might be less than 20 homologous interface templates onto which the interaction can be reliably modelled.

		Proteins in complex		
		3	4	5
Templates per interaction	1	8	64	1,024
	2	64	4,096	1,048,576
	3	512	262,144	1,073,741,824
	4	4,096	16,777,216	1,099,511,627,776
	5	32,768	1,073,741,824	1,125,899,906,842,620
	6	262,144	68,719,476,736	1,152,921,504,606,850,000
	7	2,097,152	4,398,046,511,104	1,180,591,620,717,410,000,000
	8	16,777,216	281,474,976,710,656	1,208,925,819,614,630,000,000,000
	9	134,217,728	18,014,398,509,482,000	1,237,940,039,285,380,000,000,000,000
	10	1,073,741,824	1,152,921,504,606,850,000	1,267,650,600,228,230,000,000,000,000,000

Fig 2.2: Number of possible complex models

Templates per interaction identifies how many alternative conformations are considered for each potential protein-protein interaction. Even with only five alternatives per interaction and four proteins in a complex, there are billions of possible models. We consider up to 20 interfaces per interaction.

2.3.2 Traversing an interaction network

In order to efficiently consider as many models as possible, we attempt to identity the most promising models before they are built. These are assumed to be those made up of the most promising interfaces. As the interface templates have been ranked by their similarity, we can process the most similar ones first. We start with the highest-scoring interface and assume first that it makes up part of a model. Then we recursively process the remainder of the set of interfaces. Then we assume that the first interface is not part of any model and again recursively process the remaining set of interfaces. It can be proven that this procedure iterates through the power set of interfaces, i.e. it processes every possible combination of interface templates and, therefore, every possible sub-complex topology.

This approach is preferable to traditional graph algorithms, such as the minimum spanning tree, as it does not try to find the minimal set of interactions required to hold a complex together, but rather finds the maximal set of interactions that are compatible in a complex (Fig 2.3: Traversing an interaction network). This means that it also finds ring topologies, which no existing algorithm can, and which are very common in nature, due to the fact that the evolution of protein complexes is strongly influenced by duplication events (Levy et al. 2008).

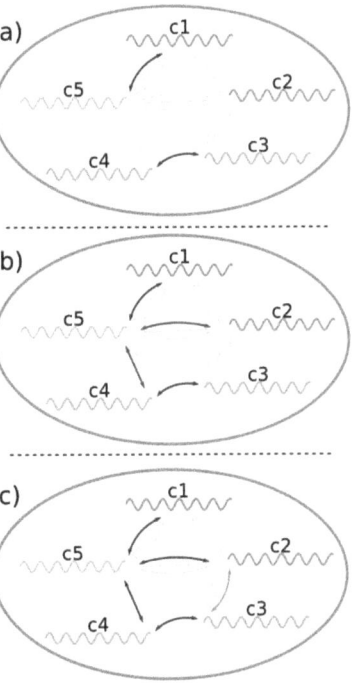

Fig 2.3: Traversing an interaction network
a) Independently added interactions (black) cannot clash. b) New interfaces (red) adjacent to existing ones (black) require clash checking. c) Both halves of a new interaction (green) already modelled, requires a check for a ring closure.

2.3.3 Merging complexes with shared components

Every interface template added to a model incorporates structural representations for additional components of the complex. In the simplest case, two dimeric complexes are merged into a single trimeric complex by superposition, i.e. structural alignment, of the protein that they both have in common (Aloy et al. 2005). In general, either one or both of the components represented by an interface template are already represented in the growing complex model. Adding the interface therefore requires structural superposition of common components.

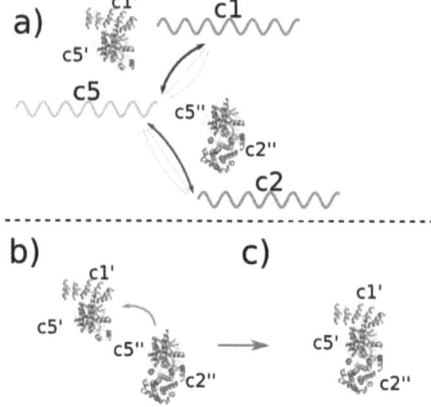

Fig 2.4: Merging complexes with shared components
a) Components c1 and c5 have been modelled (on c1' and c5' respectively), when an interface templates for c5--c2 is considered (from c5''--c2'').
b) The two templates contain a homologous component (c5' and c5''). c) The homologous structures are superposed, thereby adding c2'' to the growing complex.

For example (Fig 2.4: Merging complexes with shared components), when the first interface template c1'--c5' is added to the empty complex, it is implicitly allowed, since there are no other components present with which it could clash. Then, c5''-c2'' is added, where c5'' is a homologue of c5'. In order to orient c2'' relative to the complex c1'--c5', we superpose c5'' onto c5', which results in c2'' being placed into the common frame of reference. Then we have the trimer: c1'--c5'--c2''. The interface between c5''-c2'' is real, but the interface in the final model at that location is actually c5'--c2'' (not c5''-c2''), which is hypothetical and based on the homology between c5' and c5''.

The quality of each such superposition is a measure for how well the diverse interface templates fit together. This is incorporated into the final score for each model (Section 2.4 Scoring modelled complexes). If two interface templates, c1'--c5' and c5'--c2' are taken from the same source structure, no superposition is needed, since the c5' in each case refers to the same structure. Such events receive the maximum possible superposition score (a STAMP Sc score of 10.0 out of 10.0) to reflect that we have found a native multimeric template.

2.3.4 Identifying exclusive interactions

Interactions cannot be added to a complex model indefinitely. Indeed, the majority of these will result in steric clashes, with two proteins occupying the same space at the same time. This was the original motivation for the automated algorithm, to filter out the impossible models, leaving only those models that are internally consistent, without any structural clashes. For a potential interface template, we first check if that interaction has already been explicitly modelled on another interface template in the current complex and skip the interface in that case, considering it again in a subsequent round. Then we check whether the proteins modelled into the complex by the new interface clash with those already present in the model.

2.3.5 Detecting collisions

We identify steric clashes of a newly added interaction by approximating each protein's volume by a sphere. The sphere is located at the centre-of-mass of the protein and its radius is the radius of gyration of the protein (Fig 2.5: Collision detection). The radius of gyration is calculated by a weighted average of the distance of each atom from the centre-of-mass. The weights are defined by the atomic weights. This represents the rotational centre-of-mass of a protein. We use this as a first approximation for clash detection. As the centres-of-mass of all the interface templates in our interaction database do not change, these are pre-computed, permitting a very fast clash detection algorithm. This approximation is orders of magnitude faster than an all-atom clash detection between all pairs of proteins.

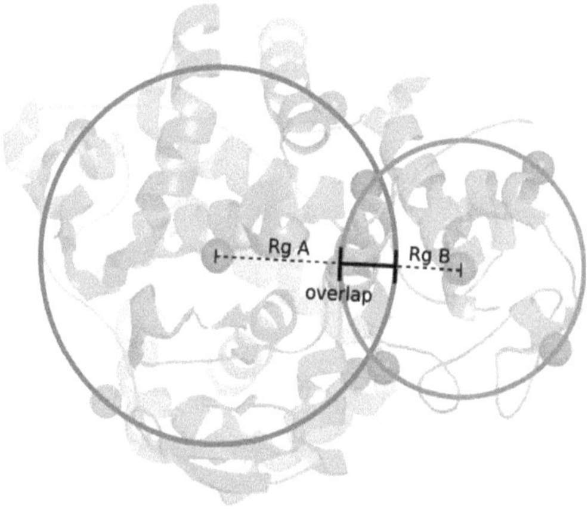

Note that we do not take the common approach of trying to identify overlapping binding sites (Ozawa et al. 2010). Firstly, because this is not unambiguously defined and, secondly, because overlapping is not black-and-white as there are cases of slightly overlapping binding sites that do not prevent simultaneous interactions. Ultimately, the question is whether two interactions can occur at the same time, which is prevented when the interaction partners try to occupy the same space at the same time, which is what we measure, though at low-resolution using spherical representations in the first instance.

Fig 2.5: Collision detection
Each protein is represented by a sphere, centred at its centre-of-mass, with a radius defined by the radius of gyration of its atoms. The red dots represent the centres of mass and the radii of gyration, extended in each direction. The overlap, and potential clash, of two proteins is measured by the overlap of these spheres. Naturally interacting proteins will overlap to some extent. This is considered a clash when the overlap is more 50% of the diameter of the smaller protein. This heuristic eliminates the majority of severe clashes.

Once a clash is found, the sub-complex being built no longer needs to be considered. This means that any components that could have been added to this sub-complex model no longer need to be considered and that search path through the interaction network can be abandoned, saving significant time that would have been wasted if we had first built every possible complex model independently. This is the most important step of the algorithm and this is what distinguishes our approach from brute-force checking of all possible models. This is the reason that clash checking must occur after each newly added interaction in a model. Indeed, it is not computationally possible to first generate all theoretically possible models, except for the most trivial complexes. Nevertheless, we also want to be certain that our spherical approximation to clash checking is still correct at the atomic level. For this reason, we perform a more precise second level of clash detection, after a model has been completed (Section 2.6 Filtering steric clashes).

The approach not only allows us to identify clashing proteins but also proteins that are in contact, when the overlap is more than zero, but does not exceed the threshold. If the contact is between two proteins with no interface templates, we may have a candidate for a novel interaction (Section 8.1.7 Novel interaction candidates). If there is an interface template for a contact, we can verify that it is consistent. In either case, we have identified a ring topology.

2.3.6 Detecting ring topologies

Unlike some previous approaches that used a minimum spanning tree algorithm (Inbar et al. 2003; Pichaud 2008), our approach is able to identify ring topologies in complexes. A spanning tree approach seeks to find the minimal number of interactions that will hold a complex together. By definition, such a complex cannot display any explicit ring topology, as a ring closure is considered to be a superfluous interaction by the algorithm. As we know that ring topologies are very common in nature (Fig 2.6: Number of ring topologies in protein complexes), due to duplication events that drive the evolution of complexes (Levy et al. 2008), it is important to identify such topologies in our modelled complexes.

Ring detection means identifying cycles in the interaction network of a single complex. Every complex of N components must contain at least $N-1$ interactions, otherwise the complex would not be connected. For every additional interface, an additional ring exists, e.g. a complex with $N+3$ interactions contains 4 rings. These may be merely rings of 3 components, where the centre of the ring is not open. However, even in this case, it is important to emphasise that a ring structure is significantly more stable than a simple structure with no ring closure. Therefore, we identify and consider such rings in our model scoring procedure as well (Section 2.4 Scoring modelled complexes).

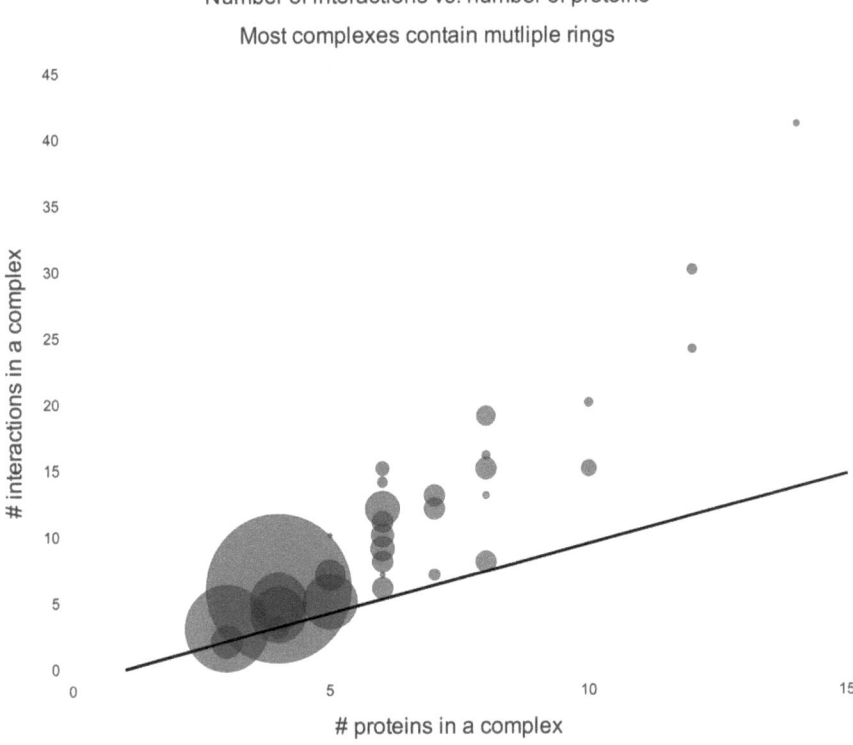

Fig 2.6: Number of ring topologies in protein complexes
Complexes are taken from the test set of benchmark structures (Section 4 Benchmark results). The line represents complexes with no rings, i.e. trimers with only two interactions, tetramers with only three. Only two data points lie on this line (3,2 and 4,3) and the majority are not only above, but show the existence of multiple rings. The size of a data point represent how often that data point occurs, e.g. complexes with four component proteins and six interactions (suggesting three rings) are most common in this set.

There are two cases of ring closure. In the first, a contact, that does not exceed the clash threshold, is identified during collision detection (Section 2.3.5 Detecting collisions) for which no known interface template exists. As no interfaces templates exist, but the components appear to be in contact, this is a candidate for a potentially novel interaction. In the second case, the growing complex model is extended by an interaction between two proteins that already have a fixed orientation to one another. In this case the interface being added serves as an additional validation of that orientation, if it fits. For example, assume a model contains component A connected to B, which is connected to C, which is connected to

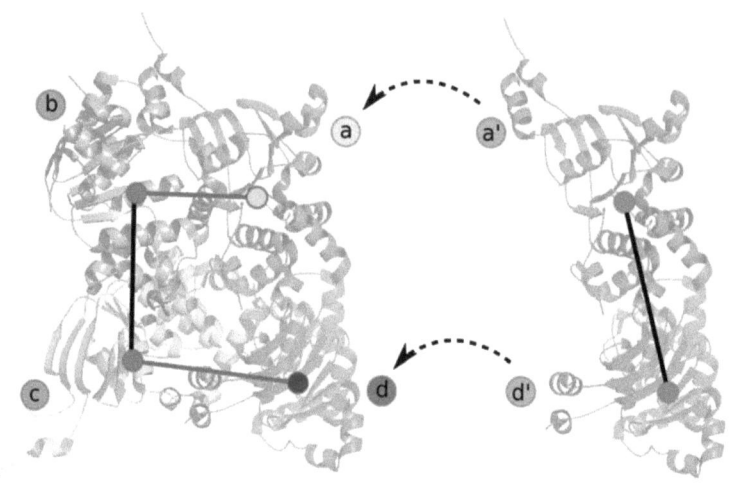

Fig 2.7: Ring detection

The tetrameric model contains three explicit interactions. The b--c (black) is from a native structure. The b--a (grey) is hypothesised on the basis of structural homology, as is c--d (grey). The *a* and *d* are simply in proximity, as a side effect of the complex assembly algorithm. However, if any known interface, e.g. a'--d' (black) can be found that can be superimposed onto the pair of *a* and *d*, without modifying their orientation, this provides structural evidence for the a--d interaction. If no such interface exists, then a--d may be a candidate for a potentially novel interface.

D, in that order (Fig 2.7: Ring detection). If an interface template exists for the interaction A--D and the interaction orientation defined by that template does not disrupt the existing complex model, i.e. is is consistent with the existing orientations of A and D, then this lends substantial credibility to the model, because the ultimate interface provides a confirmation of the appropriateness of all the previous interface templates. However, if no such interface template can fit, no ring is created. This means either that the ring has become distorted and that the model is incorrect, or that the interface may be a potentially novel interaction type. The former case can be checked by considering whether any other models contain an interface for the interaction of interest. The latter case can be verified experimentally, or corroborated by the existence of non-structural interaction data. An interface is considered to fit when the iRMSD (Aloy et al. 2003) of the interface to the iRMSD of the components closing the ring is less than 15Å.

2.4 Scoring modelled complexes

The motivation for scoring models is twofold: first, we want a relative measure to be able to rank the better models first; second, we want an absolute measure, to be able to reject poor models. There a number of relevant characteristics of a complex model that contribute, to some extent, to the final score. As noted above, our network traversal algorithm is driven by the weight of the interface templates in the network. Each model will represent some subset of these interface templates. However, a complex model is more than the sum of a set of interface templates. How the templates are related to one another, via structural alignment, measures how well the templates fit together. Whole-complex characteristics are also relevant to the quality of the final model (Fig 2.8: Scorable characteristics of complexes). As the model contains multiple proteins and multiple interfaces, many characteristics are, in fact, a list of values, one for each protein or interface. For each such list, we consider its minimum, median, and maximum as separate characteristics. This allows us to identify the best, the worst, and the common case.

Complex characteristics
Percent of components modelled
Number of source structures providing templates [1]
Number of interactions modelled
Percent of interactions modelled
Number of residues modelled
Percent of residues modelled (from all target sequences)
Globularity [2]
Buried surface area in model (ΔSAS) [3]
Percent of atoms clashing in model (within 2Å)
Interface characteristics
Interface weight (ranking of interfaces used by traversal) [4]
Number of residues interacting at interface (mean) [5]
Protein characteristics
STAMP superposition score between homologous components
Percent sequence identity
Percent of residues modelled (per component sequence) [6]

Fig 2.8: Scorable characteristics of complexes

Each characteristic contributes to the final model reliability score, each having some weight. 1) number of PDB structures that contributed one ore mote interface templates to the model. 2) the ratio of the radius of gyration to the maximum atomic radius of the complex. 3) the difference in solvent accessible surface between the complex as a whole and the sum of the solvent accessible surfaces of the monomeric proteins. 4) Section 2.2.3 Scoring interface templates. 5) E.g. when 15 residues of one interact with 10 residues of another protein, we count 12.5 residues. 6) as a percent of the residues in the one component protein.

When benchmarking (Chapter 3), we use the root mean square deviation (RMSD) between a target and a model to measure their structural difference. Our goal was to develop a model score that approximates this RMSD, and predicts it in cases when the native structure is not known. We use machine learning, via ordinary least squares (OLS), to automatically determine linear weights for the model characteristics. This defines a linear function, the value of which is an approximation of the RMSD between the model and the native structure. The weights are presented with the benchmark results (Section 4.6 Weights of model characteristics).

2.5 Clustering redundant models

Even though interface templates have already been clustered, this clustering only groups the most identical interfaces. Complete complex models may still be redundant. One reason for this is the naturally high prevalence (30% of complexes) of sub-unit duplication, i.e. internal symmetry (Taverner et al. 2008). We use an approach from computer vision, known as geometric hashing, that has previously been applied to protein structures (Nussinov & Wolfson 1991), in order to identify when two three-dimensional objects are equivalent, independent of their orientation in space. We developed our own implementation of this algorithm in the context of multimeric complexes. It identifies the location and orientation of the centre-of-mass of each component of a complex model and quickly and efficiently determines whether any previous model showed the same profile, i.e. with the same components at the same locations in space and with the same relative orientations, within a threshold of 2Å. If the components display the same relative orientations between complex models, then the interfaces used are also similar. We identify the structural class of each new model as soon as it is built. If it is new class, that model is the representative of the class. However, if it belongs to an existing class and scores better than any previous model in that class, it becomes the new representative of the class. Only the best model per structural class is reported.

2.6 Filtering steric clashes

The collision detection performed during the traversal of the interaction network (Section 2.3.5 Detecting collisions) is highly efficient and serves to short-circuit many implausible search paths. After each model is complete, we perform a more refined, and time-consuming, atomic-level check for clashes between the template structures in the complex scaffold. We use the VMD platform (Humphrey et al. 1996) to determine when more than 2% of atoms are closer than 2Å to atoms from other proteins, and discard those models in that case. This does not guarantee that the model is biochemically feasible, but it eliminates the majority of structural clashes. An atomic-level decision on the quality of a model is only possible after performing homology modelling of the target sequences onto the template structures (Section 7.6 Interaction conservation).

3 Benchmarking modelled complexes

In order to evaluate the applicability and accuracy of the complex modelling approach, it must be evaluated against complexes of known structure. These should be challenging enough to test the scope of the method on non-trivial complexes, but they must also be such that a correct complex can even be modelled with the available interface templates; all interaction templates should not come from the same structure, but there should be compatible templates for all interactions. It is not trivial to identify buildable complexes. We can make an assumption based on the availability of homologous templates, but knowing that something is buildable with a set of templates, would require doing complex modelling on it first, which is what is to be tested in the first place. We use an indirect approach to estimate when a complex should be buildable.

3.1 Defining a non-trivial benchmark

The interaction database presented above (Section 2.1 Structured interaction database) was used to initially gather a set of benchmark complexes. Each protein component was required to be structurally homologous (STAMP $Sc \leq 2.0$) to a protein in other known structures, though not necessarily all in a single complex. Likewise, each interaction in a benchmark complex was required to be homologous (iRMSD ≤ 5.0) to an interaction in other known structures, though not necessarily all present together in a single complex. This suggests that the pieces of a benchmark complex have all occurred in other known structures, which means that templates exists for building that benchmark complex.

We filtered out dimeric complexes, as these do not test the ability of the method to combine interaction data from multiple templates. We also removed nucleic acids from complexes. However, many complexes were not connected without the presence of the nucleic acid. We then removed any complexes containing any nucleic acids (Section 7.9 Nucleic acids). We use the biological units from the PDB (unpublished, ftp://ftp.wwpdb.org/pub/pdb/data/biounit) to identify and remove crystal contacts. We then removed redundant complexes, by visual inspection of the complexes.

Target complexes were then subjected to the interface template search (Section 2.2 Structured interaction networks), providing a structured interaction network for the components of the network. This template search was restricted to a maximum sequence identity of 75% to prevent finding templates from the benchmark complex itself, or from nearly identical structures. When this structured interaction network consisted of more than one disconnected sub-network (i.e. when not all the components were in the same network), these target complexes were filtered out of the benchmark dataset. This left 485 target complexes for the benchmark set.

We then determined the input sequences (as the modelling does not require that the structure of the components be known) by mapping PDB protein chains to the full-length native sequences from UniProt (The UniProt Consortium 2009). The sequence of the crystallised structure often does not correspond to the full-length native sequence. A sequence may have been truncated to aid in crystallisation and even without truncation, disordered and highly flexible segments of a protein will not be a part of the resolved structure. We therefore avoided bias by not using the sequence of the crystallised structure. The resulting complexes are shown in Fig 3.1: Sizes of benchmark complexes, Fig 3.2: Classifications of benchmark complexes, and Fig 3.3: Genera of benchmark complexes.

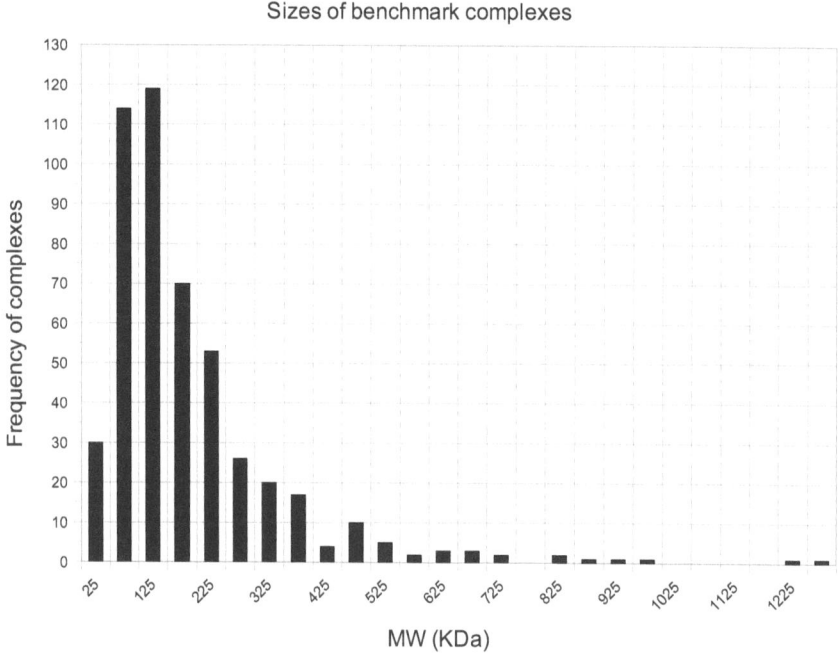

Fig 3.1: Sizes of benchmark complexes
Measured over all chains of all 485 benchmark complexes.

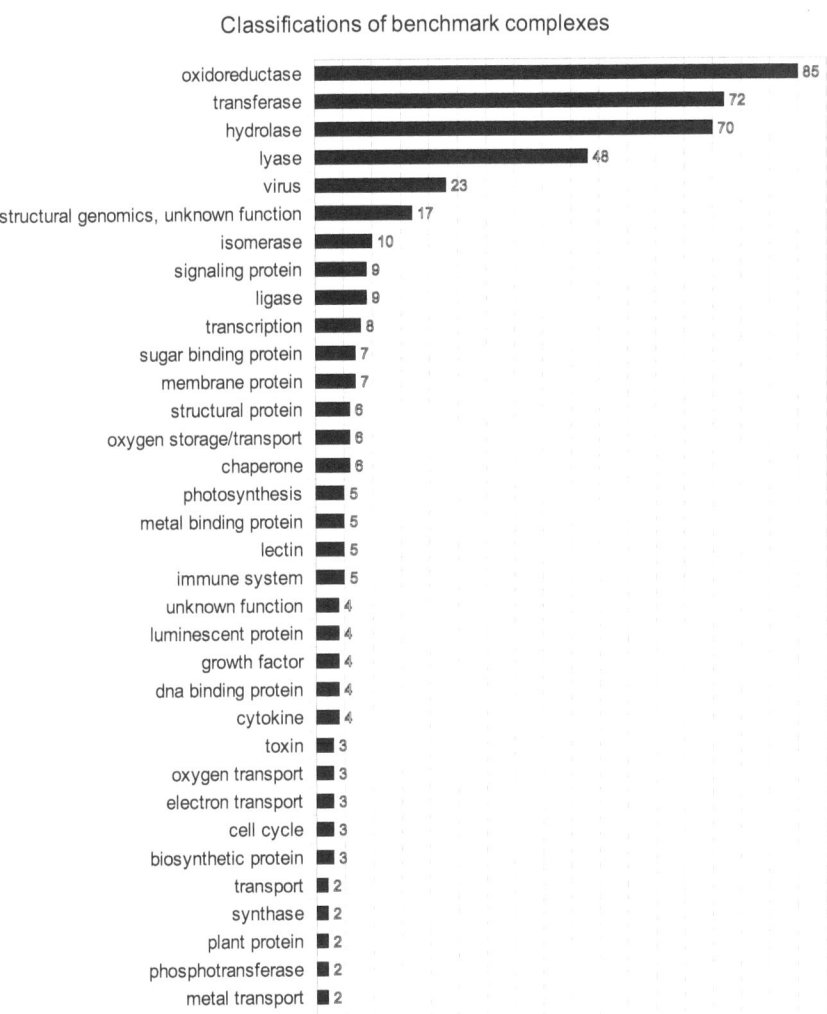

Fig 3.2: Classifications of benchmark complexes
Only classes with more than one complex are shown. Classifications assigned by the PDB.

3.1 Defining a non-trivial benchmark

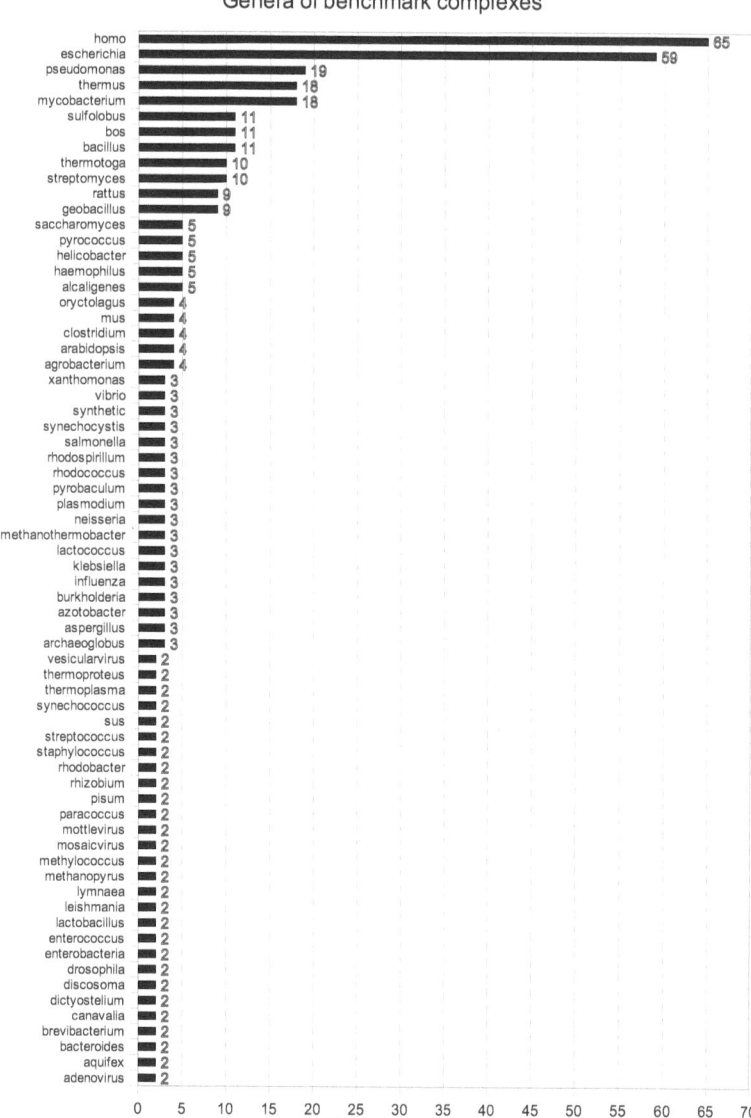

Fig 3.3: Genera of benchmark complexes
Only genera with more than one complex are shown.

3.2 Comparing a model to a benchmark complex

In order to be able to assess to what extent a modelled complex is correct, there must be a way to gauge it against the known benchmark structure. The standard measure for comparing structures is the root mean square deviation (RMSD). The RMSD measures the difference in the locations of the atoms in one structure versus another, once an atom-to-atom correspondence between the two structures (via a structural alignment) has been determined.

Like many of the points addressed above, the RMSD measure also cannot be directly extrapolated from the monomeric to the multimeric case without additional developments. Indeed, as far as we are aware, no method has been developed that can superpose multimeric complexes, except by considering each multimeric complex to be a single protein chain. This, however, can only work if the two complexes contain the same number of proteins, in the same order, in each structure. However, the assignment of protein chains in structures is arbitrary. Models also often contain only a subset of the proteins in a large interaction network. More importantly, however, complexes possess internal symmetry, whereby certain proteins are homologous to other proteins in the complex. Before any RMSD can be measured, a correspondence between the proteins of two

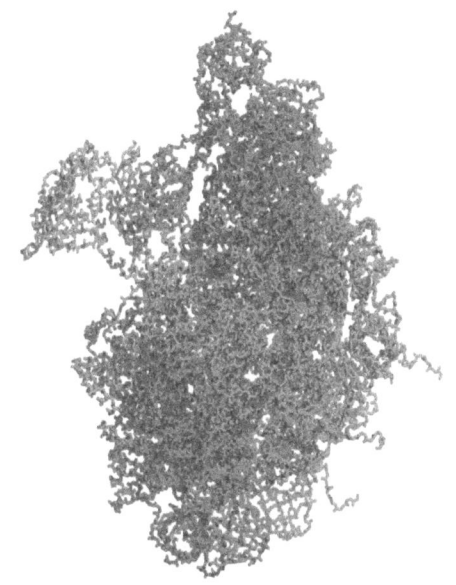

Fig 3.4: Misalignment of complexes resulting from internal homology

Bovine F1-ATPase (PDB ID 1E79) contains a hexameric ring around the main stalk. This ring contains three alpha and thee beta subunits, which alternate around the ring. As each alpha subunit is equivalent, superimposing the wrong alpha subunit from the model results in the wrong superimposition of the whole complex. Even though this model can be visually verified to be correct, standard structural alignment methods provide no way to measure this.

complexes must be found. The existence of internal homologies in each of the complexes multiplies the number of possible associations between two complexes. In the case of F1-ATPase (Fig 3.4: Misalignment of complexes resulting from internal homology) there are two groups of three identical proteins in the hexameric ring. When a model also contains this complete ring, there are

$$\frac{3!}{(3-3)!} \times \frac{3!}{(3-3)!} = 36$$

possible ways to generate a correspondence between the two rings. This number is substantially larger when the model does not contain every protein in the target. In general, each of these possibilities must be tested.

We have developed a heuristic algorithm that explores these possibilities, looking for the best fit based on multiple possible mappings between proteins. We first identify classes of homologous proteins within a complex via all-against-all pairwise sequence alignments, using ClustalW (Thompson et al. 1994). This is followed by single-linkage clustering of pairs of proteins that are 90% identical. Note that knowing the stoichiometry of the components in a complex is not sufficient here, as this only identifies 100% identical components. Rather, we find all internal homology by identifying also highly similar components, because these may be modelled using the same templates. Once the homologous classes within each of two complexes have been identified, the correspondence of classes between the two complexes must be determined. In benchmarking, however, this is already known, as the model represents the same classes as the benchmark complex. Therefore, this step is necessary when comparing two complexes in general, but can be skipped when comparing a modelled complex to its benchmark. The correspondence between classes then leads to a list of mappings of proteins between the two complexes. For each one of these mapping we can measure the RMSD for the entire complex. We identity the "correct" mapping as the one that leads to the smallest RMSD. Of course, this is not necessarily a good RMSD, which would indicate that the model in question fails the benchmark.

For each mapping between complexes, we first sequence align corresponding proteins. Note that a structural alignment would be more accurate than a sequence alignment here, but we avoid bias by using a sequence alignment when evaluating models against benchmark complexes. This is because it is the alignment that is the most decisive factor for the quality of homology models and we assume that the target structure is unknown while benchmarking. The sequence coordinates from the alignments must then be translated to residue coordinates of the structures. This mapping is provided by our interaction database (Section 2.1 Structured interaction database). Gap positions in the alignment, as well residues that have no structure are removed from the alignment. This results in a mapping between the residues of two components that are to be compared. This is done over all component proteins, providing a complex-level mapping. This is then used to calculate the optimal superposition of the C-alpha atoms between two complexes, using least-squares fitting (Kabsch 1976). With this superposition, the RMSD can then be measured over all of the aligned C-alpha atoms between the two complexes. This procedures is repeated for every correspondence of protein chains between two complexes. Generally, many of the mappings will be incorrect and some mappings may lead to equivalent complexes, due to internal symmetries. Therefore, heuristics are used to reduce the computational complexity of the problem, as it is often not feasible to go through the entire list of mappings. We abandon the search if the RMSD does not improve by 1% in 1000 steps.

3.3 Avoiding parameter bias

Weights for the model characteristics presented above (Section 2.4 Scoring modelled complexes) are learned from known structures. Deriving these weights from a set of complexes and evaluating our performance on the same set would be a case of parameter bias (Varma & Simon 2006). We avoid this by learning the weights on a randomly determined subset of 50% of the complexes, the training set. When then apply these weights to the remaining 50% of the complexes, the test test, to evaluate our performance (Chapter 4 Benchmark results). Thus, no complex in the set of complexes being tested had any influence on the weights being used during benchmarking. To verify whether these scores correspond to the actual quality of the models, we evaluated each model from the test set against its respective benchmark complex using the C-alpha RMSD between them.

After the results on the test set were produced and evaluated, the method was re-trained for general applicability using the weights from the entire set of benchmark complexes, both the training and the test set. Thus, the performance reported on the test set, based on weights from the training set, will be reflective of the performance of the method on unknown complexes, when using weights from the entire benchmark set (Neto & Dougherty 2004).

4 Benchmark results

4.1 RMSD threshold for correctness

Having a method for computing the RMSD between a modelled complex and a benchmark structure does not tell us at what RMSD threshold a model is "correct". We do not expect the RMSD to reach zero, due to the natural flexibility of proteins (Burra et al. 2009; Hasegawa & Holm 2009). For our purposes, a model is correct when it maintains the overall shape, the locations of the individual proteins, the shape of the individual proteins, the relative orientations between the proteins, i.e. the interface orientations, and the general secondary structure. We checked this visually on the models produced before concluding that an RMSD threshold of 10Å best fulfil these criteria (Fig 4.1: Interpreting RMSD between complexes).

We found that an RMSD below 2Å identifies an identical complex, i.e. the complex is of known structure or can be trivially assembled. At 3Å there may be individual loop shifts, though the secondary structure is conserved. At 5Å loop shifts are larger. At 10Å slight domain shifts begin to occur, though the relative orientations between the components is retained, which retains most, but not all interfaces. At 15Å domain shifts become pronounced, details of individual domains are no longer reliable, and interfaces are lost. At 20Å the complex, and maybe the individual proteins, may occupy the same volume, but domains and interfaces are not structurally comparable.

4.2 Modelling coverage

Starting with 591 target complexes, 485 produced structured interaction networks with the potential to model complexes of at least trimeric size. Models of varying correctness could be constructed for 418 of these. The targets that produced no models were not considered and the remaining targets were randomly split into training and test sets (209 targets each). Limiting the number of models per target to 50, the training set resulting in 5673 total models; the test set contained 5288 models. Of the test targets, 156 (75% of 209) produced at least one correct model (RMSD ≤ 10Å). Of these, 156 (100%) produced at least one correct model that had also been predicted to be correct (Fig 4.2: Coverage of benchmark set of complexes). In other words, if a target has correct models, we are successful at identifying them as such. (Note, models do not necessarily provide a structure for every protein component in a target, but are always at least trimeric).

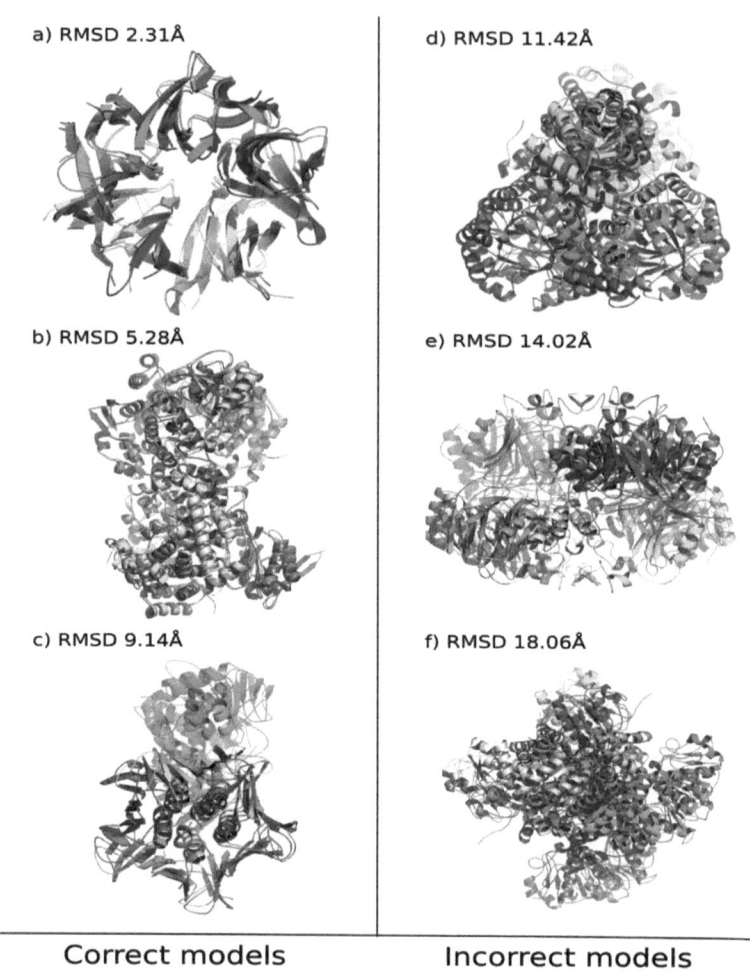

Fig 4.1: Interpreting RMSD between complexes

a) Nucleoplasmin from Xenopus laevis (2vtx), reconstructed with 2.31Å difference from the model. Slight loop rearrangements are possible. b) Diol dehydratase-cyanocobalamin from Klebsiella oxytoca (1egm), at 5.28Å from the model. Slight domain shifts are possible. c) Shiga toxin from Shigella dysenteriae (1dm0), at 9.14Å from the model. Four of five domains superpose well, while the fifth (top) displays a significant shift; the overall model is still correct, however. d) 2,3-dimethylmalat lyase from Aspergillus niger (3fa3), at 11.42Å. Every domain shows a significant shift. The model cannot be considered correct, though the overall shape and the shape and orientation of the domains are correct. e) Homoprotocatechuate 2,3-dioxygenase from Brevibacterium fuscum (3eck), at 14.02Å. Three of four domains show significant shifts, the fourth (bottom right) is completely wrong and simply occupies the same volume as the native protein. f) Malate oxidoreductase from Thermotoga maritima (2hae), at 18.06Å. Domain shift is so severe that the correspondence between components is difficult to identify visually.

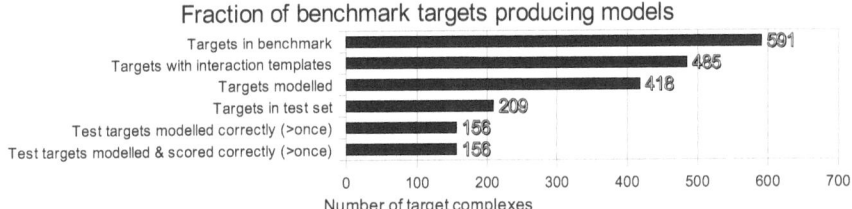

Fig 4.2: Coverage of benchmark set of complexes

A correct target is one with at least one model with RMSD ≤ 10Å. A target is modelled and scored correctly when the score is also ≤ 10Å. This shows that a target that is modellable is also identified as such. Here the size (i.e. fraction of proteins modelled) of the complexes is not considered.

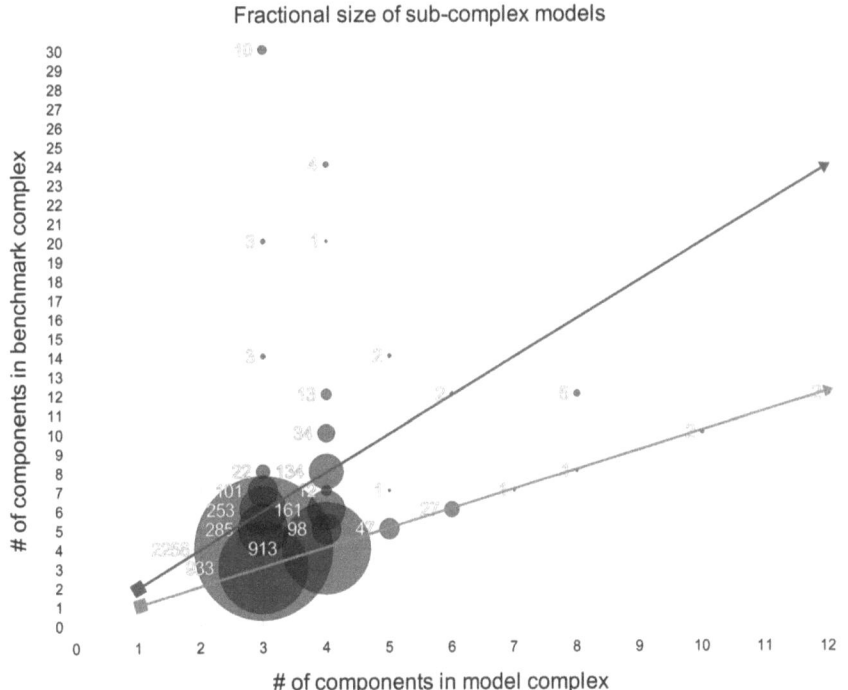

Fig 4.3: Size of sub-complex models

Component proteins in models, as a fraction of the number of proteins in the benchmark complex. For example, the largest point occurs at 3,4 showing that in 2258 models, 3 out of 4 components were modelled. For each benchmark complex, up to 50 models may be represented. In the majority of benchmark targets, between 50% (red line) and 100% (green line) are modelled.

4.2 Modelling coverage

In addition to the fraction of test targets modelled, coverage can also be assessed at the level of individual complexes, by the fraction of proteins modelled (Fig 4.3: Size of sub-complex models). We are not able to produce full-size models for many targets, but succeed in modelling at least 50% of the complex in most cases. This is largely dependent on the stringency of our clash detection algorithm (Section 2.3.5 Detecting collisions), which is optimised for the most common cases.

We then considered only the full-size models, i.e. where 100% of the components are present in the model, and identified the best model in each case (according to our model score). The frequency of these best scores (Fig 4.4: Best-scoring complete models) shows a number of complexes within our score threshold (≤ 10.0), thought the majority lie in the grey area with a score between 10 and 15. This is one reason why we accept sub-complex models. More importantly, however, the extent of a complex, in terms of which proteins are involved in it, can generally be an overestimate, if it is extracted from a noisy interaction network.

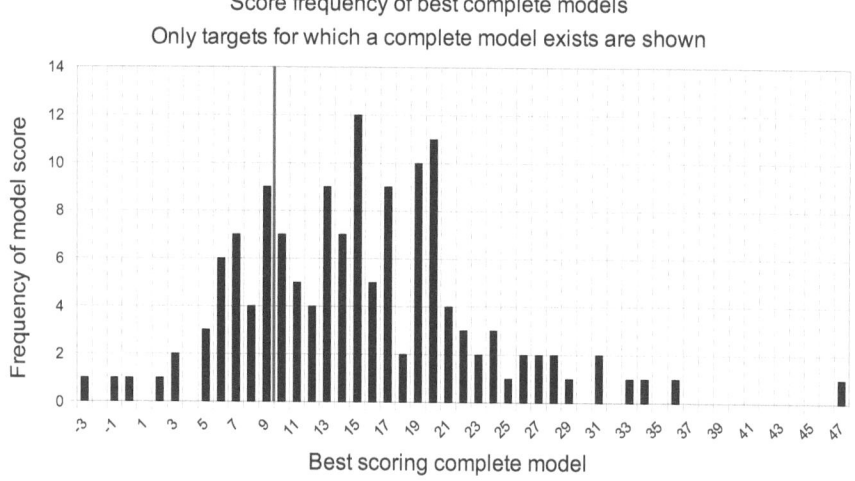

Fig 4.4: Best-scoring complete models

For each target in the benchmark, the best score of the models with 100% of the protein components is considered. This shows that the majority of the complete models lie beyond the 10Å threshold (red line). In 42 cases there is a full-size model that falls within the threshold. If we relax this to 15Å, we can get 37 more full-size models, but this is a grey area. Sizes of the individual complexes not shown.

4.3 Accuracy

The correlation between our subjective score and the objective RMSD over all models in the test set is 0.55 (Fig 4.5: Correlation between backbone RMSD and prediction score). The majority of models are correctly identified as true negatives. The false negatives (upper left quadrant) largely outweigh the false positive (lower right quadrant), reflecting the high specificity of our scoring. While there is no perfect linear correlation between our learned score and the RMSD, it is interesting to note the slight gap (decreased density of points) along the vertical line around RMSD=10.0Å. There is a large density of models below 5Å and above 15Å. This provides additional support for our threshold (RMSD ≤ 10.0Å) for deciding when a

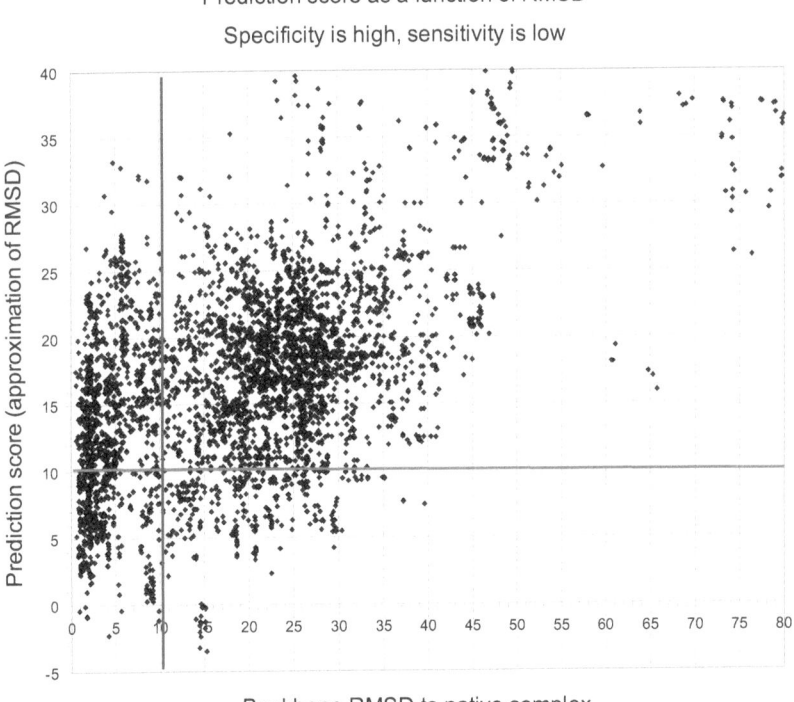

Fig 4.5: Correlation between backbone RMSD and prediction score
Correlation: 0.55. The red vertical line marks the RMSD threshold; all models to the left are correct. The horizontal green line marks the prediction threshold; all models below are predicted to be correct. (The linear function is an approximation and can produce scores < 0. These are also positive predictions). The lower left quadrant shows the true positives (TP), the lower right the false positives (FP), the upper left the false negatives (FN), and the upper right the true negatives (TN).

4.3 Accuracy

model is correct (Section 4.1 RMSD threshold for correctness). As our score aims to approximate the RMSD, we also set our decision threshold for a positive prediction to $score \leq 10.0$. This cut-off produces the contingency matrix shown in Fig 4.7: Contingency matrix. To consider the effect on performance at different score thresholds, we considered the ROC curve (Zweig & Campbell 1993), which shows the trade off between sensitivity and specificity Fig 4.6: ROC curve). An ideal ROC curve approaches the upper left corner of the plot. Here,

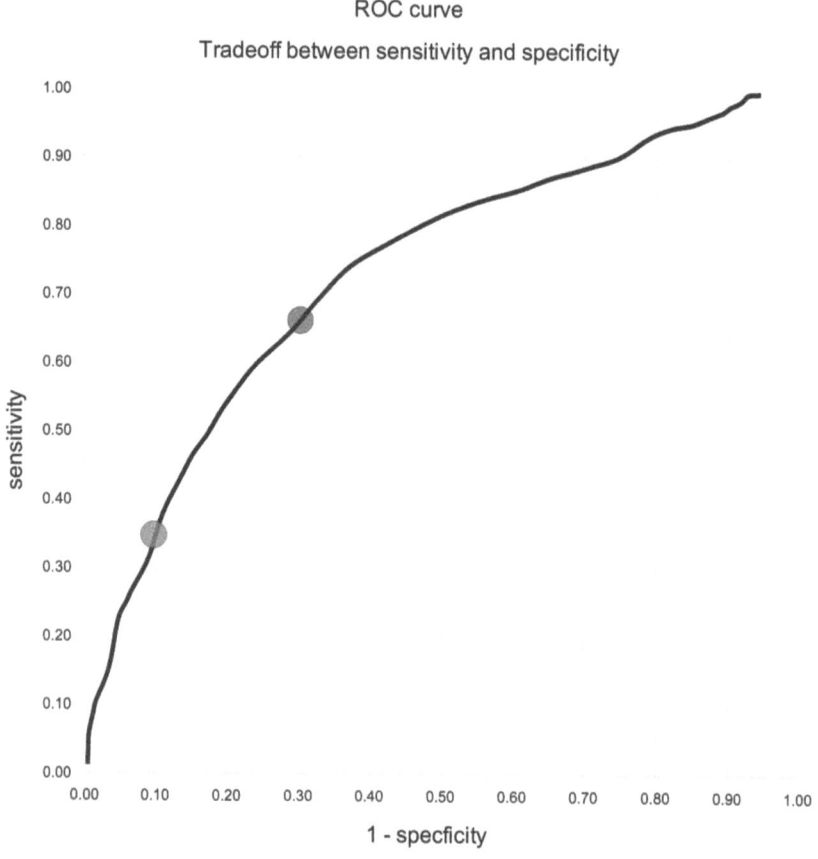

Fig 4.6: ROC curve

The receiver-operating characteristic (ROC) curve (Zweig & Campbell 1993) shows the relationship between sensitivity and specificity at different decision thresholds. The threshold for a model being correct is fixed at RMSD≤10Å, while the threshold for the score is varied. A low score threshold is strict and maximises specificity, at the cost of sensitivity. A high score threshold is lenient and maximises sensitivity at the cost of specificity. The default threshold of $score \leq 10.0$ corresponds to a specificity of 0.90 and a sensitivity of 0.36 (green dot). The red dot identifies the optimal trade-off.

the point along the curve closest to the upper left point of the plot (red dot) has a specificity near 0.65 (1 - 0.45) and a sensitivity near 0.75. However, our goal is not necessarily to achieve optimal balance, as the cost of false positives (believing that a wrong model is right) is higher than the cost of false negatives (believing that a right model is wrong). Hence, we prefer the more stringent threshold (green dot), which increases confidence in the best scoring complexes, at the cost of increased false negatives.

	$score \leqslant 10$	$score > 10$
$RMSD \leqslant 10 \text{Å}$	(TP) 182 (3%)	(FN) 1536 (29%)
$RMSD > 10 \text{Å}$	(FP) 107 (2%)	(TN) 3458 (65%)

Fig 4.7: Contingency matrix
A correct model has RMSD ≤ 10Å. A positive prediction is made for models with a *score* ≤ 10.

4.4 Ranking models

In order to evaluate whether the heuristics of the traversal algorithm (2.3.2 Traversing an interaction network) succeed in finding the most promising models first, we considered how often the Xth model generated was the best model for the target (Fig 4.8: Rank of the best-scoring model per benchmark target). In more than half of the cases (121 of 209) the best model occurs within the first five models produced.

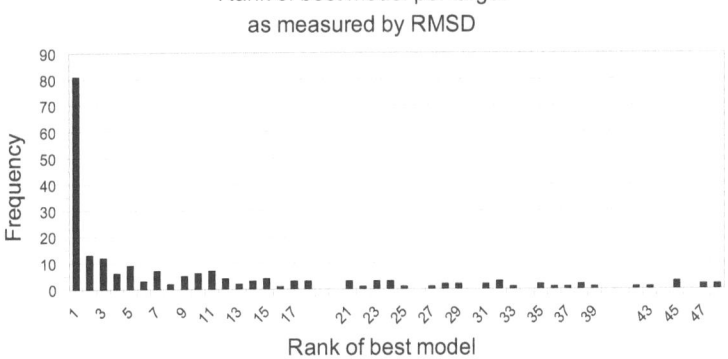

Fig 4.8: Rank of the best-scoring model per benchmark target
Best is defined in terms of the objective RMSD, not the subjective model score. The search heuristic leads to the best model being produced earlier in the search. The fraction of target proteins modelled in each complex is not considered here.

4.5 Sequence identity threshold of modellability

Considering the sequence identity threshold on modellability (Fig 4.9: Sequence identity limits of modelling), we see that correct models (RMSD ≤ 10.0) can be produced with sequence identities as low as 25%. This is consistent with previously published limits for homology modelling (1.4 Modelling interfaces). High sequence identity alone cannot guarantee a correct model, however, as noted by the incorrect models with 75% sequence identity. Highly homologous sequences such as these are not difficult to model as monomers (Fig 1.4: Applications of (monomeric) homology models at various levels of accuracy). However, multimeric homology modelling is more than getting the individual components right. The reason these complexes are wrong is due to the differing interface orientations, which cause individual protein components to be shifted.

Fig 4.9: Sequence identity limits of modelling

Below 25% sequence identity, homology modelling of interactions, and therefore, complexes, is generally not possible, consistent with previous results (Section 1.4 Modelling interfaces). All models with RMSD ≤ 10.0Å are considered correct (red line). Note that maximum sequence identity was limited to ≤ 75% on this test set. The fraction of protein components modelled in each case is not shown.

4.6 Weights of model characteristics

After benchmarking the test set, using the weights derived from the training set, we derived another set of weights using both the training and test sets. The performance on the test set using these weights then reflects the expected performance on previously unseen complexes. In addition, the final weights can also be used to identify the most significant characteristics of complex models (Fig 4.10: Score weighting determined by ordinary least squares (OLS)). There are many characteristics of low significance. Two stand out, however. The median STAMP superposition score, which measures the structural homology of shared components, is the most determining factor for the quality of a model. On the other hand, the number of interactions modelled is the largest determining factor of a poor model. This is because larger complexes are more difficult to model. Note that there are some side effects here as well. For example, that larger complexes are more difficult is expected, but it is not immediately clear why the 'Percentage of components modelled' appears significant. This may reflect the fact that a complete model is often a sign of finding a near-native structure as a template, and not that complete models are more likely to be correct, which we have shown is generally not the case (Fig 4.4: Best-scoring complete models). Also that the maximum STAMP superposition score is negatively correlated seems to be a side effect. With multimeric (trimeric or larger) templates, no superposition is required to combine interfaces, these are assigned a maximum STAMP score (10.0). In other words, these are cases where a full multimeric complex was used as a template. If the sequence identity is poor, however, one full

Description	Weight
STAMP superposition score (median)	-2.35
Number of source structures	-0.48
Interface weight (median)	-0.29
Percent buried surface area	-0.27
Percent of components modelled	-0.25
Interface weight (min)	-0.22
Percent of residues modelled (complex)	-0.10
Globularity	-0.06
Number of residues interacting at interface (min)	-0.06
Percent of residues modelled (protein) (max)	-0.02
Number of residues interacting at interface (max)	-0.02
Percent sequence identity (min)	0.02
Number of residues modelled (complex)	0.01
Percent sequence identity (max)	0.01
Percent of residues modelled (protein) (min)	0.01
Number of residues interacting at interface (median)	0.02
Interface weight (max)	0.04
Percent of residues modelled (protein) (median)	0.10
Percent of interactions modelled	0.16
Percent of atoms clashing in model	0.18
Percent sequence identity (median)	0.19
STAMP superposition score (max)	1.43
Number of interactions modelled	2.80
Constant (from OLS fitting)	55.10

Fig 4.10: Score weighting determined by ordinary least squares (OLS).

Each characteristic contributes to the final model reliability score, which intends to approximate the RMSD to the native structure. Larger absolute values have a larger significance. More negative values lead to a better score (green), more positive values lead to a worse score (red).

complex is a poor model for another. In that case, better models can be made by combining better templates from different structures. This would agree with the fact that the number of source structures is relatively correlated to the prediction score. It is also interesting to note that the median interface weight is significant, whereas the maximum interface weight is not. This is intuitive, since even a poor model may have some good interfaces, but this is no guarantee that the whole model is correct. The interface weight is the one characteristic that guides the network traversal (Section 2.3.2 Traversing an interaction network), which explains why our traversal algorithm succeeds in finding the best models early (Fig 4.8: Rank of the best-scoring model per benchmark target).

5 Defining the yeast complexome

The approach was separately applied to two sets of yeast complexes. The first includes a set of complexes from a whole-proteome study of *S. cerevisiae* using tandem affinity purification and mass spectrometry (TAP MS) (Gavin et al. 2006). Complexes were defined by iterative clustering on the socio-affinity score between proteins, which is a measure of the propensity of proteins to interact, determined from multiple rounds of TAP MS (Fig 5.1: Socio-affinities between complex components). These complexes were merged with related studies on yeast (Gavin et al. 2002; Aloy et al. 2004) by 3D Repertoire (www.3drepertoire.org), where we retrieved the data. This provided a set of 615 complexes, from which we considered the 508 that contained at least three protein components. The average complex contained 13.6 proteins (Fig 5.3: Sizes of 3D Repertoire complexes).

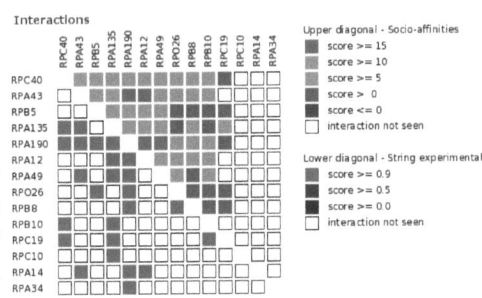

Fig 5.1: Socio-affinities between complex components RNA polymerase I (ID 154B). From 3DRepertoire.org.

We additionally used a set of 408 low-throughput yeast complexes from CYC2008 (Pu et al. 2007; Pu et al. 2009) These are manually curated from small-scale studies of specific yeast interactions and complexes. This provided a set of 408 complexes, of which we considered the 236 that were at least trimeric. The average complex contained 4.7 proteins (Fig 5.4: Sizes of CYC2008 complexes). As with the 3D Repertoire complexes, no explicit stoichiometry of the components was available in this set.

5.1 Structured interface templates

In addition to the interaction templates automatically derived from the Protein Data Bank (PDB) (Section 2.1 Structured interaction database), we derived homologous interface templates specifically for all yeast interactions (Mosca et al. 2010, in preparation). We first defined the yeast proteome using the Saccharomyces Genome Database (Cherry et al. 1998) (status October, 2009). We extracted all open reading frames (ORFs), mapped these to unique UniProt (v15.8) entries (The UniProt Consortium 2009). This provided 5817 proteins.

5.1 Structured interface templates

First, we identified interfaces from yeast complexes of known structure, as these require no modelling. A known structure is defined as having a Blast E-value ≤ 10E-4, sequence identity ≥ 98%, alignment coverage ≥ 80% and at least 8 residue-residue interactions between chains in the biological unit (i.e. quaternary structure) of the structure. This produced 5937 interfaces, covering 463 unique interactions. Second, we added full-chain interfaces from homologous interactions (interologues). We required sequence identity ≥ 30%, alignment coverage ≥ 90%. This produced 31605 interfaces, covering 978 unique interactions. Third, we added domain-domain interactions from homologous structures, using 3DID (Stein et al. 2005). We required sequence identity ≥ 30% and alignment coverage of a known PFAM (Bateman et al. 2004) domain of ≥ 90%. This produced 2181176 interfaces, covering 2497 unique interactions. Fourth, we included templates applicable to yeast from the automatically generated interface database described above (Section 2.1 Structured interaction database). This produced 17416 interfaces, covering 1617 unique interactions. Fifth, we added templates obtained from high-throughput docking of all yeast proteins (Mosca et al. 2009). Up to three docking poses were considered for each potential interaction. These were ranked by their pyDock score (Cheng et al. 2007). After filtering based on the docking threshold from the original study (pyDock score ≥ 1386), this provided 1792 templates, covering 1639 unique interactions. In total, these five categories provided 2237926 potential interface templates (Fig 5.2: Structured interaction network of S. cerevisiae). As these may overlap, the total number of unique interactions covered by this set is smaller than the sum of the unique interactions from each set (Fig 5.5:Sources of yeast interface templates).

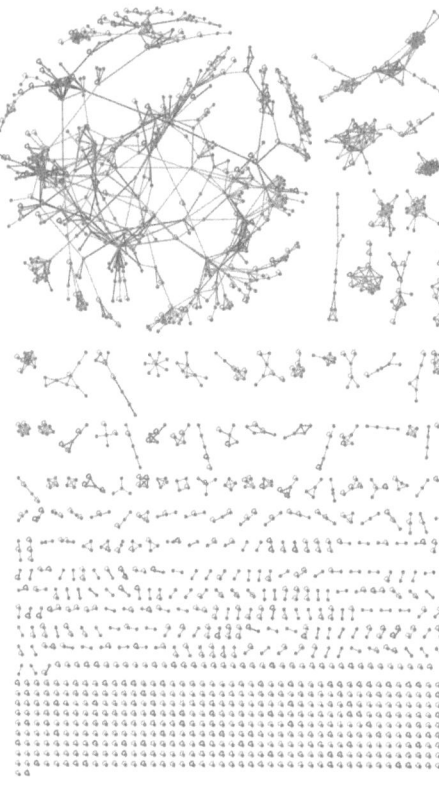

Fig 5.2: Structured interaction network of *S. cerevisiae*. Yeast interactions supported by known structures. Single nodes with self loops represent homodimers.

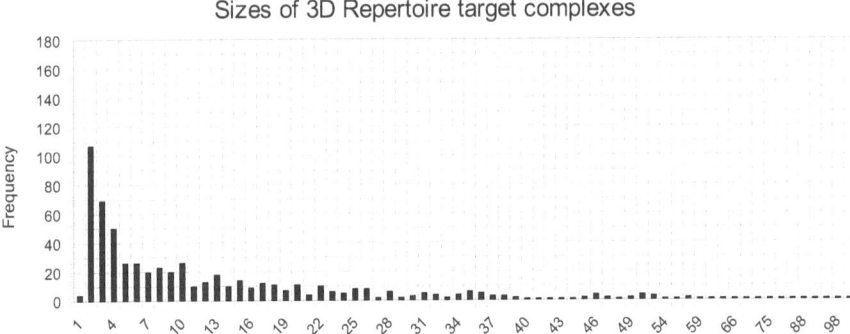

Fig 5.3: Sizes of 3D Repertoire complexes
Median number of proteins per complex: 8, average: 13.6

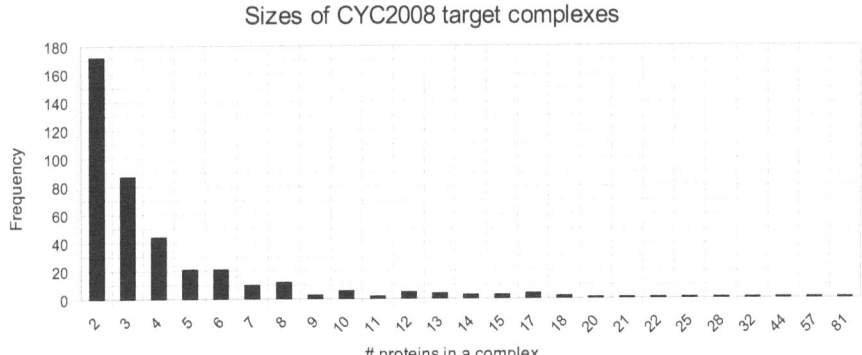

Fig 5.4: Sizes of CYC2008 complexes
Median number of proteins per complex: 3, average: 4.7

Type	Interfaces	% of total interfaces	Unique interactions
Known structure	5937	0.27%	463
Full-chain interologues	31605	1.41%	978
Domain-domain interologues	2181176	97.46%	2497
Interaction database	17416	0.78%	1617
Docking	1792	0.08%	1639
Total	**2237926**	**100.00%**	

Fig 5.5: Sources of yeast interface templates

For a potentially interacting pair of proteins in a complex, we considered the 20 top-scoring homologous templates. Only when there were less than 20 such templates did we consider the docking templates. This is because our interface scoring system (Section 2.2.3 Scoring interface templates) is not compatible with the docking scoring system, as discussed above (Section 1.4 Modelling interfaces). The network of structured interaction templates shows two extremes. On the one hand, many isolated dimers are not able to be linked to any other complexes. On the other hand, there is a large cluster of interactions with many overlaps and no clear complex boundaries (Fig 5.2: Structured interaction network of S. cerevisiae).

5.2 Structure of individual components

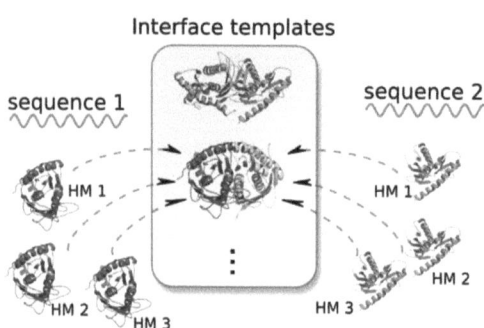

Fig 5.6: Using known structures and homology models
For each interface template, the native structure, when known, or up to the best three homology models is superimposed onto the corresponding half of the template, before complex assembly begins. HM: homology model.

While our complex modelling procedure does not require that the structural of individual monomers be known, we used the known structure, when available, to improve the quality of the clash checking during assembly. In cases where a close homologue of known structure was available, we used homology models, obtained from ModBASE (Pieper et al. 2006; Pieper et al. 2008). We required a ModBASE score ≥ 0.7 (an internal metric reflecting overall model quality), sequence identity ≥ 30%, and alignment coverage ≥ 90%. Whether a native structure or homology model is used as a structural representative, it is superposed onto each of the interface templates of each of the interactions that a protein potentially participates in. In other words, each interface template is replaced by the best possible structural representative of a pair of proteins, where the orientation is defined by the original interface template. This is done for each interface template for each pair of components in a complex. This also has the advantage that the complex models produced are not merely scaffolds, but contain the actual sequences being assembled. In cases where no homology model was available, we used the structure from the interface template as the structural representative, as in the standard version of the method (2.3.3 Merging complexes with shared components). Docking interfaces also made use of the homology models. Native-native dockings made up 9.5% of the docking templates, native-model made up 22.8% and model-model made up the remaining 67.8% of the docking templates.

6 Results of yeast complex modellings

Overall we produce similar numbers of models for the 3D Repertoire (39) and the CYC2008 (36) datasets. Roughly half of these, in each case, represent complexes for which the structure, or a part of the structure has already been resolved. In the CYC2008 case, however, this represents 38% (36 of 94) of the complexes for which we identified any homologous templates, whereas this is only 12% (39 of 326) of the 3D Repertoire complexes that had homologous templates. This reflects the fact that the latter is a whole proteome study, containing many more unannotated complexes. We use the known structures as a validation of the application of our approach, then we look at some examples of predicted complexes.

6.1 3D Repertoire coverage

Of the 508 multimeric complexes from 3D Repertoire, we could sufficient templates to proceed with modelling for 326. We produced models for 109 target complexes, having 2433 models total (maximum 50 models per target). We then removed those models with excessive clashes (more than 0.10% of atoms within 2Å of another atom) and those whose median sequence identity (across all modelled components) was less than 30%, which also excluded any models assembled solely from docking templates. This resulted in plausible models for 39 target complexes Fig 6.1: Modelling coverage of 3D Repertoire complexes. That slightly less than half of the multimeric complexes did not generate interaction networks shows that many interactions are not yet modellable on the available structural templates. For those with templates, several were rejected after determining that the interaction templates were not compatible with one another structurally. Complexes for which plausible models could be produced are shown in Fig 6.2: Modelled complexes: 3D Repertoire. Some of these are covered by known yeast structures, as shown by a very high median percent sequence identity in those cases. In other cases, however, the native structure is not identified, due to the approximate clash detection, which erroneously filters some of these out. Relaxing the

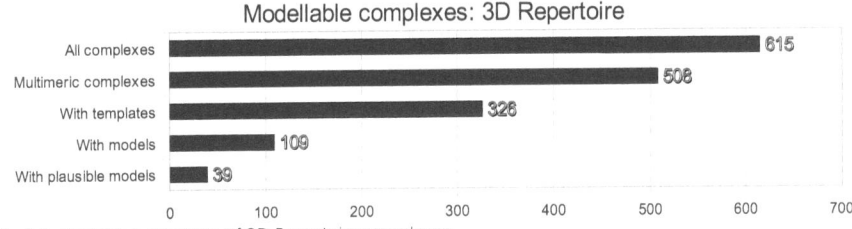

Fig 6.1: Modelling coverage of 3D Repertoire complexes

6.1 3D Repertoire coverage

Acc.	Complex	Proteins	Modelled	Plausible	Median % ID
106	20S core particle of the proteasome	45	14	14	100.00
148	19S regulatory proteasome subcomplex + associated regulatory complex	39	11	9	100.00
151	unknown	45	22	20	100.00
157	19/22S regulator	43	12	10	100.00
589	unknown	11	6	50	100.00
838	unknown	18	4	50	100.00
858	ISW1/IOC3 complex	37	3	1	100.00
130	unknown	15	5	50	99.69
219	DNA polymearse alpha-primase complex	15	4	48	99.69
840	Ctf18/Rfc2/Rfc3/Rfc4/Rfc5 complex	19	5	50	99.69
139	RNA polymerase II mediator complex, SRB subcomplex of RNA polymerase II	64	3	3	97.60
235	Nup84 sub-complex	23	6	22	95.44
725	Coatomer COPII complex	13	4	9	95.44
143	unknown	25	4	19	92.66
857	unknown	24	3	2	92.66
086	RSC, abundant chromatin remodeling complex	50	4	2	92.34
102	TRAPP I complex, TRAPPII complex	13	6	50	91.67
500	Exosome 3'-5' exoribonuclease complex	36	8	48	88.21
154B	RNA polymerase I	14	7	6	82.62
655	Protein phosphatase 2A complex	20	6	30	74.03
839	unknown	12	4	2	67.21
154A	RNA polymerase III	38	12	18	67.14
113	RNA polymerase II-associated Paf1 complex	15	4	21	62.11
070	unknown	21	3	1	60.80
582	unknown	45	4	12	60.80
859	Casein kinase II complex	25	6	32	60.80
727	unknown	10	3	9	56.90
850	Pab1/eIF4G/eIF4E complex	40	5	28	56.78
096	unknown	6	4	26	52.69
032	unknown	13	4	6	50.73
083	unknown	10	3	6	50.63
165	Cyclin-dependent protein kinase complex	10	3	6	50.63
509	RNA polymerase I	18	5	12	50.57
156	unknown	103	4	1	50.36
847	Actin-associated motorproteins 1 complex	12	5	23	47.54
030	inactive PKA holoenzyme, cAMP-dependent protein kinase complex	4	4	15	43.34
687	unknown	8	3	1	38.19
079	septin ring	7	4	50	37.71
101	Translational release factor complex	30	6	42	36.38

Fig 6.2: Modelled complexes: 3D Repertoire

Plausible: number of alternative complex models retained. Modelled: Number of proteins in the largest model. Median % ID: Median sequence identity of all modelled proteins in the best model. The best model and the largest model are generally not the same model.

threshold does identify more of the known structures (data not shown), but has the side effect that it dramatically increases the number of clashes in all of the other models (data not shown). Our clash threshold is a compromise between these. A number of the 3D Repertoire complexes are of unknown function (15 of the 39 that have plausible models, Fig 6.2: Modelled complexes: 3D Repertoire). These models may be able to provide insights by identifying the sources of the interaction templates going into the models. This is complicated, however, by the large size of some of the complexes, suggesting that multiple complexes have been merged, making it challenging to identify specific functions. This will require that we first break down the defined complexes into functional modules and identity for which of those plausible models are possible (Section 7.2 Complex composition).

6.2 CYC2008 coverage

Of 236 multimeric targets, for 94 targets we identified enough templates for modelling. For 36 of those targets, models were produced, with 955 individual models total. After applying the same filtering that we applied to the 3D Repertoire models, we still had at least one plausible model for each of those 36 targets (Fig 6.3: Modelling coverage of CYC2008 complexes). A larger portion of the CYC2008 complexes lead to plausible models, compared to the 3D Repertoire set. One reason for this is that the CYC2008 complexes are considerably smaller (Fig 5.4: Sizes of CYC2008 complexes). Another reason is that the CYC2008 models used none of the docking templates. The docking templates are most useful for adding components from low-confidence interactions, whereas the complexes of CYC2008 are generally composed of higher-confidence interactions. We also found that the docking interactions had a slightly increased likelihood of producing clashes when integrated into a modelled complex (data now shown). As with the 3D Repertoire complexes, the most homologous model and the largest model are generally not the same model. Our scoring system ranks the models by their estimated accuracy (Section 4.4 Ranking models), but this does not consider how large the model is, though the larger models are often of greater interest. There is no integrated way to rank the complexes, as there is a trade-off between quality and quantity (of components in a complex model). We have ranked the models by median sequence identity simply to identify the cases where we are dealing with a complex of known structure (Fig 6.4: Modelled complexes: CYC2008).

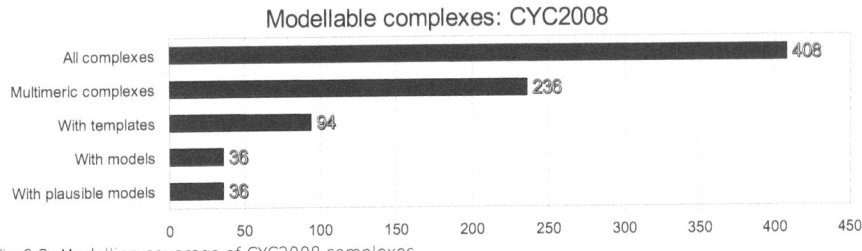

Fig 6.3: Modelling coverage of CYC2008 complexes

Acc.	Complex	Proteins	Modelled	Plausible	Median % ID
120	ESCRT-II	3	3	13	100.00
190	methionyl glutamyl tRNA synthetase complex	3	3	1	100.00
119	ESCRT I	4	4	4	100.00
250	Nucleosomal protein complex	8	8	2	100.00
78	Cytochrome bc1	10	9	6	100.00
99	DNA-directed RNA polymerase II	12	12	9	100.00
4	20S proteasome	14	14	9	100.00
124	F0/F1 ATP synthase	18	5	50	100.00
83	cytoplasmic ribosomal small subunit	57	4	7	100.00
82	cytoplasmic ribosomal large subunit	81	7	23	100.00
95	DNA replication factor C (Elg1p)	5	4	50	99.69
96	DNA replication factor C (Rad24p)	5	4	50	99.69
97	DNA replication factor C (Rcf1p)	5	5	50	99.69
94	DNA replication factor C (Ctf18p/Ctf8p/dcc1p)	7	4	50	99.69
350	Snf1p/Snf4p/Sip2p	3	3	5	98.41
70	COPII	11	4	8	98.20
176	mediator (SRB) complex	25	3	2	98.09
385	TRAPP	10	4	40	91.67
348	Snf1p/Snf4p/Gal83p	3	3	7	77.78
349	Snf1p/Snf4p/Sip1p	3	3	5	77.78
23	Arp2/3	7	5	50	69.28
285	protein phosphatase type 2A (Rts1p)	3	3	4	60.00
34	calcineurin	3	3	7	58.97
393	U5 snRNP	14	7	50	58.33
390	U1 snRNP	17	8	50	58.33
391	U2 snRNP	18	7	50	58.33
67	commitment complex	21	8	50	58.33
284	protein phosphatase type 2A (Cdc55p)	3	3	3	56.00
203	mitochondrial ribosomal small subunit	32	4	50	55.00
392	U4/U6 x U5 tri-snRNP	28	9	50	51.45
35	cAMP-dependent protein kinase	4	4	50	51.26
106	eEF1	5	3	11	51.26
17	AP-1 adaptor	5	3	17	47.99
79	Cytochrome c oxidase (complex IV)	11	5	50	46.15
3	19/22S regulator	22	7	50	45.96
100	DNA-directed RNA polymerase III	17	3	9	43.48

Fig 6.4: Modelled complexes: CYC2008
Plausible: number of alternative complex models retained. Modelled: Number of proteins in the largest model. Median % ID: Median sequence identity of all modelled proteins in the best model. The best model and the largest model are generally not the same model.

6.3 Reconstruction of known complexes

While we have systematically benchmarked our approach on a set of complexes of known structure (Chapter 4 Benchmark results), the set of known yeast complexes serves as a practical validation of our modelling procedure. These complexes distinguish themselves by the fact that the complex composition was not precisely defined and by the fact that the stoichiometry is generally not given. This determines whether, given native dimeric interactions, they can reasonably be fit into the complete complex. In practice, one can identify the existence of a native structure without having to reconstruct it. However, this is a real-world test that shows what we can expect from modelling when high-quality templates are available but where the composition or the stoichiometry may be incomplete.

The structure of the yeast 20S proteasome has been resolved via X-ray crystallography (Groll et al. 1997). It contains two copies of two heptameric rings, giving a total of 14 unique components. One of our models identified this native structure in its entirety. Interestingly, the 14 modelled components did not simply identify the two heptameric rings, as one might expect. Rather, one of the heptameric rings was identified completely; the second ring contained four of the seven remaining components, and the remaining three components were placed into the third ring already, adjacent to the second ring (Fig 6.5: Proteasome models). This shows that, had the stoichiometry not been known, the presence of two half rings would have suggested how the symmetric structure likely fits together, i.e. that the 4-component and 3-component rings are complementary. This also shows that the algorithm does not follow any intuitive geometrical path when assembling models, but may reveal additional details based on where components are modelled into a complex. Finally, using the correct stoichiometry, with two copies of each of the 14 components, we verified that the algorithm was able to find the full, 28-component, native structure (after running three iterative modelling rounds).

Fig 6.5: Proteasome models
Models were generated using only one copy of each of the 14 proteins. Rather than a double ring model, as one might expect, the algorithm had already begun to model pieces of the third ring, suggesting a four-ring assembly. Two such models together make up the entire proteasome.

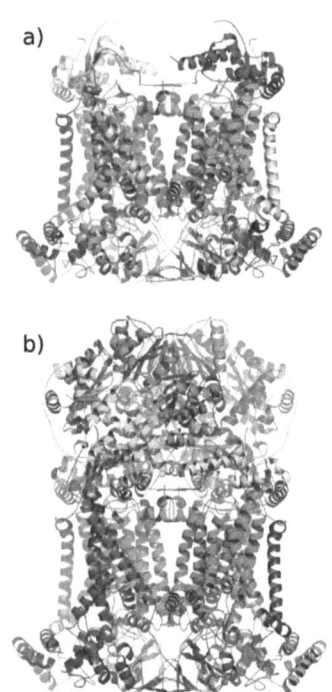

Fig 6.6: Cytochrome-bc1 (Complex III)
a) 12-component model after using 2 copies of each of the 9 components. b) Using a) as a seed, the full 18-component complex was reconstructed.

Cytochrome-bc1 (Complex III / co-enzyme Q) is a complex of 18 proteins, of which nine are unique (Hunte et al. 2000). Using only the nine unique components, we were able to identify the native structure of half of the complex. The presence of homodimeric templates may have suggested increasing the copy number of the components, one-by-one. Instead, we directly used two copies of each of the nine components. We used the 9-component complex that we already built as a seed for the subsequent round of modelling. This produced a 12-component model, which we then used as a seed for the final round of modelling, which identified the 18-component native structure (Fig 6.6: Cytochrome-bc1 (Complex III)). This iterative re-seeding with previous sub-models is necessarily for very large complexes, such as complex III, due to the number of ways that 18 proteins can be arranged in a complex. It is often the case that that intermediate models suggest what the final assembly may look like. This allows one to take a more directed path through the interaction network, once one has a specific structure in mind. We also looked at complex II (succinate dehydrogenase). While the native structure is not known, a model of the tetrameric state using homology from *E. coli,* as well as docking, has been published (Oyedotun & Lemire 2004). None of our models met our quality requirements for this complex, however.

While the structure of yeast RNA polymerase II is known (Darst 1991), our complexes have been defined by the clustering from the 3D Repertoire interactions. This cluster contained only nine of the twelve core components. The composition of RNA polymerase II had been determined before the structure, however (Myer & Young 1998), and this was also reflected in the corresponding CYC2008 complex. Merging these two provided the full set of 12 components. Given all of the components, we were able to model between 10 and 11 components in most cases. In the end, we again relaxed the clash threshold from 50% to 75%

overlap, which finally allowed us to reconstruct the full 12-component native structure. We do not use this threshold of 75% in general, however, due to the amount of excessive clashes that this tolerates in other complexes. This shows that it is necessary to adjust one's approach based on what can be identified from intermediate complex models. We also looked at RNA polymerase I (14 components) and RNA polymerase III (17 components) which share a common core with RNA polymerase II. The core contains the five components: RPB5, RPB6 (i.e. RPO26), RPB8, RPB10 and RPB12 (i.e. RPC10). We were not able to extend RNA polymerase I beyond this common core. For RNA polymerase III we were able to model up to 12 components, but closer inspection revealed that many models had been disrupted by too many poor docking interfaces. In neither of these two cases was the native structure found, showing that simply relaxing the clash threshold is not a universal solution.

For F1 ATP synthase (unpublished, PDB ID 2WPD) we also had to merge the 3D Repertoire and CYC2008 sets, before we were able to find a sub-complex of the native structure containing the central stalk and most of the surrounding rotor. Again this required adjusting the clash threshold from 50% to 75% due to the non-globular nature of the components in the stalk (Fig 3.4: Misalignment of complexes resulting from internal homology). The shows again that non-globular components are generally more challenging for our simplified representation of components (Section 7.5 Clash detection).

We looked at multiple variants of the DNA replication factor c (RFC) complex: Ctf18p/Ctf8p/dcc1p variant (seven components), Elg1p (five components), Rad24p (five components), and Rcf1p (five components). In the first three cases, we were able to identify four components each. In the last case (Rcf1p), we identify all five components from the native structure. These models all find their templates from one structure structure (Bowman et al. 2004), showing that we are able to identify the common core, but lack templates that are homologous enough to make a refined distinction between the different complex variants. This shows that the resolution of our models are dependent on the level of diversity among the homologous templates that are available. Different complexes can be modelled to different extents, depending on well the individual components have been studied.

The native structure of the nucleosome core particle (White et al. 2001) shows a homotetramer of intertwined heterodimers. This again required using a more relaxed clash threshold of 75% to be able to model the intertwined heterodimeric unit, which is then structurally repeated for the other three units, ultimately reconstructing the native structure.

We had expected that the exosome structure from human (Liu et al. 2006) would have sufficed to model the yeast exosome, as others have suggested (Taverner et al. 2008). Using the components given by 3D Repertoire, we were able to generate models of up to eight components, though none of these identified the core hexameric ring. A multitude of docking templates appear to have increased the number of clashes, preventing us from finding the native structure. The fact that the complex was described with 36 potentially interacting components also seems to have led to much time being spent modelling less confident interactions. Given only the six components of the core ring, however, the native structure is easily identified and reconstructed.

We reconstructed the native structure of the tetrameric ESCRT I complex (Kostelansky et al. 2007). For the ESCRT II complex, the three components in the set were able to identify the native structure (Teo et al. 2004), which contains four components, as it has two copies of VPS25. This shows that considering the source of the template structures can provide important contextual information about the complexes we are modelling. This underscores the importance of looking at each source structure, particularly when trying to infer the stoichiometry of the components. We had no results on ESCRT III, for which no structure nor homologous structures appear to be known.

SNF1 (yeast homologue of AMPK / AMP-activated protein kinase) is of known structure (Amodeo et al. 2007) and all three components were correctly identified. Two other complex variants included the components sip1p and gal83. As these are each homologous to sip2p of SNF1, we used the same yeast SNF1 to also model these two other variants. As we saw above with DNA replication factor c, the diversity of homologous templates is not always sufficient to differentiate between close variants such as this, but provides at least a common scaffold, with which other studies can examine how particular residue differences might cause deviations from the common model.

6.4 Predicted complex structures

The definitions of the complexes between the two data sets overlap, with the CYC2008 more often representing the core of the complex and the 3DR Repertoire including more, and potentially novel, interaction partners. In some cases, we complemented a 3D Repertoire complex by adding any missing components listed in the corresponding CYC2008 complex. This was on a case-by-case basis. In both sets of complexes, we were missing data on the stoichiometry of the components for most of the complexes. The models generally have one copy of each component. In some cases, literature evidence provides the correct stoichiometry, for some or all of the components.

While we were able to produce models for the ribsomal small (CYC2008 #83,#203) and large (CYC2008 #82) subunits, closer inspection revealed that the models were not well connected, due to small protein-protein interfaces. As the structure is stabilised by RNA, which we do not model, our models contain only the skeleton of the proteins, without the RNA scaffolding, which makes modelling complexes with high nucleic acid content currently beyond our reach. This would be theoretically possible, however (Section 7.9 Nucleic acids).

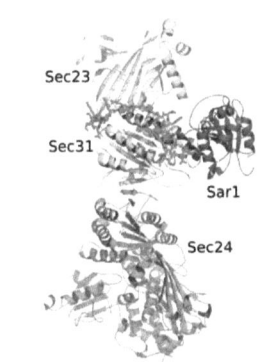

Fig 6.7: COPII model

For the COPII coatamer complex (CYC2008 #70) (Fig 6.7: COPII model) we identified two native yeast structures (Bi et al. 2002; Bi 2007), one with Sec23,Sec31,SAR1, the other with Sec23 and Sec24. We merged these on the shared Sec23 component, thereby adding Sec24 and producing a tetrameric model of high confidence. There is also a yeast structure of a sub-complex containing Sec13 and Sec31 (Fath et al. 2007). Unfortunately, we could not link these on the common Sec31 components, because the Sec31 component in the larger structure only contained a truncated 49-residue protein. For the remaining components we are still missing adequately similar template interactions. This reveals the difficulties in relying on any one structure which may not tell the whole story. This is why we consider multiple templates for each interaction.

For Arp2/3 (CYC2008 #23) (Fig 6.8: Arp2/3 model) we were able to model five of the seven identified components using various bovine templates, each of which show a 7-component structure, ranging from 31% to 74% sequence identity to the corresponding

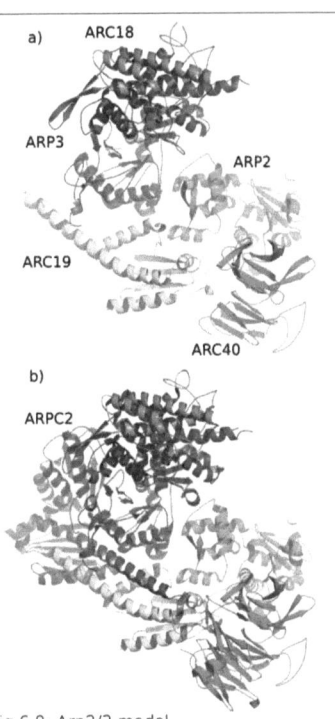

Fig 6.8: Arp2/3 model

a) Model with five components. b) Bovine template includes ARPC2, but the highly non-globular component cannot be fit into the model

yeast sequences. We were not able to include ARC35 in our model based on the multimeric bovine complex, despite its homology to ARC19, which was in our model, and despite the existence of a bovine template for the interaction (ARC35 corresponds to bovine ARPC2) (Nolen & Pollard 2007). Here also, the very non-globular shape of the components prevented us from fitting the ARC35 into the model, which represents five of the seven components from the bovine structure. Fortunately, the bovine template is similar enough, that we can anticipate how our model will look with a more refined clash detection method.

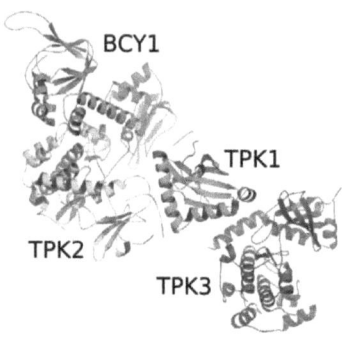

Fig 6.9: cAMP dependent protein kinase

For cAMP dependent protein kinase (PKA) (CYC2008 #35) (Fig 6.9: cAMP dependent protein kinase) we created a tetrameric model using two components from mouse PKA at 40%-54% identity (Wu et al. 2007) combined with two components from cow PKA at 51%-51% identity (Gassel et al. 2004). The TPK1--TPK3 interface is partially missing, because we were only able to fit TPK1 in by modelling it on a sub-segment template (149 of 397 residues). This is another case where a more refined clash detection would be beneficial.

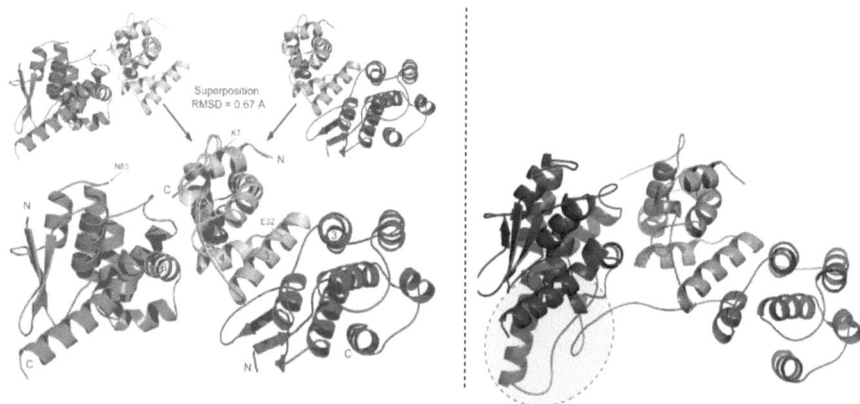

Fig 6.10: Methionyl glutamyl tRNA sythetase

Merged from two dimeric structures. The shared component (GU4, green/yellow) is identical, making the superposition trivial. However, the originally published model (Simader et al. 2006) does not show an additional helix (highlighted) that is present in the original Methionyl-tRNA synthetase (blue), which clashes slightly with Glutamyl-tRNA synthetase (red)

Our Methionyl glutamyl tRNA sythetase model (Fig 6.10: Methionyl glutamyl tRNA sythetase) would normally have been dismissed due to a clash. However, we discovered that the two templates that we used had already been superimposed into a trimeric model under the original study (Simader et al. 2006). Our model identified the same shared GU4 component for our trimeric model and put the components in the same orientation. Interestingly, the originally proposed model does not display an extra helix that is present in the Methionyl-tRNA synthetase component from the dimeric structure (Fig 6.10: Methionyl glutamyl tRNA sythetase). This helix clashes with the Glutamyl-tRNA synthetase from the other dimeric structure. There may be a solution, in that the helix is connected by a long loop that would most likely be flexible enough to accommodate another orientation. This case shows that a clash need not necessarily always invalidate a model. Though, this shows the challenges introduced by potentially flexible regions and conformational shifts in general (Section 7.8 Alternative conformations). Overall, however, this suggests possibilities for potentially larger Class I tRNA synthetases suggested by TAP MS complex discovery approaches.

The TRAPP (transport protein) complex is an eicosamer consisting of two copies of each of ten components (Sacher et al. 2000). We originally started with one of each of the ten components and were able to find a structure from yeast containing four of the ten. A number of models also incorporated TRS33, not part of the original structure, using an interaction template from an interaction between BET3 and BET5. As TRS33 is not similar to BET3, this suggests rather that BET3 may occur at this position. The interface template was part of the original structure (unpublished, PDB ID 3CUE), but was subsequently removed from the curated Biounit dataset as a suspected crystal contact. Interestingly, this supposed crystal contact positions BET3 (which we misidentified as TRS33) adjacent to the native BET3 and displays a plausible homodimeric interface (Fig 6.11: TRAPP complex extended via

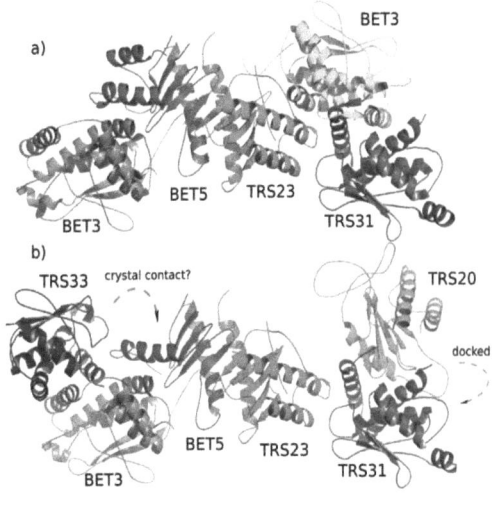

Fig 6.11: TRAPP complex extended via docking
a) Structure of yeast TRAPP
b) Model with TRS20 (cyan) docked onto TRS31 (red) in place of the original BET3. The second BET3 (blue, mislabelled as TRS33) may dimerise with the other BET3 (orange).

docking). This interface is novel and not from the original structure; it is induced by the orientation of the other interactions. However, it is consistent with the rest of the model and agrees with the published stoichiometry (Sacher et al. 2000). This suggests that the interfaces dismissed as crystal contacts may need to be reconsidered.

We were also able to include a new component, TRS20 via a high-confidence docking onto TRS31 (score 1737 > 1386). This takes the place of on instance of BET3, which as we mentioned, seems to also fit elsewhere in the complex. Though, the docked TRS20 does not preserve the contacts to TRS23 that the original BET3 had. We cannot yet make a strong conclusion on this model until we have more information on the location of the remaining components. Unfortunately, the template complex contains many ambiguous interfaces, some of which are crystal contacts. Given the similarities between BET5 and BET23 and the apparent ability for BET3 to dimerise, there are many more possibilities to consider.

These predicted complexes have shown that there many aspects that affect the quality of a model and the number of components that can reasonably be included in a model. For one, a compromise must be met, between too lenient and too strict clash thresholds. Ultimately, this demonstrates the need for a higher-resolution clash detection method (Section 7.5 Clash detection). This is a technical issue, however. More decisive is knowing which components to model into a complex and their stoichiometry. We have seen that the template structures themselves can occasionally provide hints on the stoichiometry. At the very least, a homodimeric template is a direct indication that the copy number of that component should probably be at least two. Identifying symmetry in intermediate models can provide further clues. We have also seen how modelling very large complexes often requires using an intermediate sub-complex model as a seed. This allows one to partially direct the path through the vast search space of the interaction network. This shows that complex modelling is an iterative process. Although the individual steps are fully automated, context can and should be used to decide which steps to take.

7 Discussion

Overall our benchmark proved the potential of our approach. Applying the method to real interaction data from the whole yeast proteome presented new challenges in terms of the fuzzy boundaries between complexes and due to the general lack of stoichiometry of the components. Docking templates provided less useful templates, as expected, but also led to an increase in the number of clashes in our models. Clash detection is generally an aspect that we are still improving (Section 8.1.3 Refined clash detection), though parameter adjustment still allows us to identify many complexes if we expect the components to be non-globular. The sequence identity threshold we find is consistent with previously published results (Section 1.4 Modelling interfaces), lending credibility to our models. We are not yet able to apply the method to complexes containing nucleic acids, though this is theoretically possible. We also discuss a case where we were able to determine the likely function of an unannotated complex, whose structure was determined in the context of structural genomics initiatives.

7.1 Scoring

On our benchmark set, target complexes were chosen based on the existence of potential template structures. This does not guarantee that similar structures can be found, because we imposed a sequence identity limit of 75% in an effort to make the test realistic (Section 3.1 Defining a non-trivial benchmark). The benchmark simply defines a set of target complexes for which interaction templates can be found. The existence of templates, however, does not guarantee that they will be structurally compatible within a single complex. In fact, testing this structural compatibility is the motivation for this method in the first place.

The benchmark also shows that there is a trade-off between the number of components that can be incorporated into a complex and the quality of the resulting assembly. Complete models (those containing every component protein) showed worse scores in general (Fig 4.4: Best-scoring complete models). This is why we model all possible sub-complexes and do not try to force every component into each model. This is also reflected in the learned weights of the model characteristics, where we found that the attribute most negatively associated with the RMSD was the number of interactions modelled (Fig 4.10: Score weighting determined by ordinary least squares (OLS)). This is intuitive, as modelling larger complexes is more difficult. This was also evidenced on the yeast complexes, where the majority of the scores did not fall within the threshold ($score \leq 10.0$) that we had set for our benchmark models (Section 4.1 RMSD threshold for correctness). This does not suggest that there were no plausible models. Rather, the score is a relative measure, which serves to rank the models. This ranking performed well on the benchmark, producing better models earlier during the search (Section 4.4 Ranking models). The models from the yeast complexes that we judged as plausible also had a tendency, though less pronounced, to be produced earlier in the algorithm. This association will improve once the clash checking resolution is improved (Section 8.1.3 Refined clash detection). This also shows that our scheme for scoring interfaces, though based simply on sequence identity and interface size (Section 2.2.3 Scoring interface templates), is much better than our scheme for scoring whole complexes. A complex is more than the sum of its interactions, however. This is underscored by the fact that the superposition scores (from combining dimeric templates) were found to be the characteristic most positively correlated with a model's correctness (Fig 4.10: Score weighting determined by ordinary least squares (OLS)). Some of the learned characteristics are redundant (for multi-valued characteristics we consider the minimum, median, and maximum), others are likely missing (Fig 2.8: Scorable characteristics of complexes). Our first priority has been to eliminate structurally impossible complexes, to reduce to the number of models from a googol (Fig 2.2: Number of possible complex models) to dozens (Fig 4.8: Rank of the best-scoring model per benchmark target).

In practice, the final decision on the plausibility of a model will not be based on an automated score. The score verifies that we are able to identify when a complex is modellable and when one model is preferable to another. In practice, the evaluation of a model is based on how well it fits together, the functional annotation of its template components, and the species they are derived from. It requires a holistic judgement, based on the biology of the complex being modelled. This is why some of the uncharacterised complexes from the 3D Repertoire set created a challenge and why, in some cases, we complemented the 3D Repertoire complexes with the corresponding CYC2008 complexes, to refine the composition of the complexes.

7.2 Complex composition

Knowing the composition of a complex is necessary to be able to produce reliable models, or even to identify when a complex is of known structure. As we saw with RNA polymerase II (Section 6.3 Reconstruction of known complexes), the clustering of the TAP-MS data led to some components being missing. After merging this set with the corresponding set from CYC2008, it was straight-forward to reconstruct the known structure. In general, merging corresponding sets from the two data sources helped to refine the composition of complexes, but we did not do this in every case, as increasing the number of components in a complex increases the computational complexity and decreases the number of possible models that we can check. Complexes of approximately 12 components (Fig 6.2: Modelled complexes: 3D

Repertoire, Fig 6.4: Modelled complexes: CYC2008) define our current limits. That the average 3D Repertoire complex consists of 13.6 components (Fig 5.3: Sizes of 3D Repertoire complexes) shows that it was too ambitious to model these complexes without refining the definitions of the core complexes. It will be more useful to identify the core components of the 3D Repertoire complexes, and then to also merge these core complexes with the CYC2008 set, in order to model the cores of complexes before beginning the second stage where we try to extend these cores with lower-confidence interactions (8.1 Current work). This is what enabled us to identify the cytochrome-bc1 and proteasome complexes (Section 6.3 Reconstruction of known complexes), though automating this will allow us to take short-cuts, such as beginning the search with multimeric template structures, rather than assembling models using only dimeric templates (Section 8.1.5 Hierarchical assembly).

7.3 Stoichiometry

One important reason for the increased performance on the benchmark, versus the yeast set, is that the complex composition, as well as the stoichiometry, was implicitly given. We could have removed the stoichiometry data from the components in the benchmark set, leaving just one copy of each sequence-unique component. We did not test this, since our method does not claim to be able to automatically determine stoichiometry based on structural fitting. This is also a serious challenge for EM-based based complex assembly (de Vries et al. 2010; Alber et al. 2008). Having the stoichiometry of the components makes a significant difference. We found that using the stoichiometry in the cases where it was known made identifying known structures from the yeast set much more likely (Section 6.3 Reconstruction of known complexes). While stoichiometry data is not available explicitly, the existence of a homodimeric template can also suggest a plurality of a component, though this would still not tell us the exact number of copies of each component. Short of simply putting two copies of such a component into a complex, however, it is generally not computationally feasible to systematically guess copy numbers for each component.

7.4 Docking templates

We observed that docking templates, even with high-confidence docking scores, were slightly more likely to result in clashes in the final model. We may have used docking templates too early in the algorithm. There is no way to rank the docking templates among the homology modelled templates, as they are based on different assumptions (Section 1.4 Modelling interfaces), most importantly that these are *de novo* predictions, and not based on the experimental structure of homologues. To prevent this, we used docking templates only when too few homologous templates were available. This resulted in some complexes being modelled using only docking templates. This is also problematic when ranking the complex models, as the scoring system has not been trained to accommodate docking scores, leaving no way to compare them to the homology models (Section 2.4 Scoring modelled complexes). Finally, we used only the high-confidence docking templates (Section 5.1 Structured interface templates), but this only reduced the extent of the problems and did not eliminate them. Docking templates, like homologous interface templates, can be over estimated. The benefit of our approach is that it screens for the interfaces that make structural sense in the structural context of a complex. In this light, the docking templates that remain receive more credibility, though we cannot easily quantify this. Regarding the conclusion of Section 7.2 , Complex composition, it would also make sense here to model in two phases, where docking is not considered when modelling the complex core, but added in subsequent modelling rounds when considering the lower-confidence interactions (Section 8.1.4 Defining core complexes). However, a more refined clash detection method would have been able to pre-emptively eliminate more of the clashes during the traversal algorithm itself.

7.5 Clash detection

There are two stages to clash detection. The first is performed during the assembly algorithm. It verifies that protein components do not collide when placed into a complex model. In order to be efficient, it assumes globularity of components. (Section 2.3.5 Detecting collisions). Highly non-globular components may erroneously be rejected at the default threshold (50% overlap). For this reason, the benchmark set includes complexes of mostly globular components. To test the approach also on non-globular components, we looked at a ligand-gated ion channel from *Gloebacter violaceus*. We were able to find templates at 22% sequence identity from an ion channel from *Erwinia chrysanthemi*. By relaxing the clash threshold from 50% to 75% we were able to build the complex correctly, with a C-alpha RMSD of 5.33Å (Fig 7.1: Gloebacter violaceus (GLIC) ion channel).

We had similar results on the yeast complexes, where we were able to reconstruct the known structures of F1 ATP synthase and RNA II polymerase once we had relaxed the threshold again from 50% to 75%. We do not use this relaxed threshold in general, however, as this effectively eliminates the strength of the algorithm, which is to filter out the many models that are structurally impossible. This shows that non-globular proteins are within our reach, but they may require some *a priori* knowledge. This is often clear from the initial models. If small models of very non-globular components are produced, additional modelling rounds with a looser clash threshold can help incorporate non-globular components.

The second stage clash detection uses a high-resolution atomic check for steric clashes in the final model. It is a time-consuming process that is only used as a post-filter, meaning that it can no longer improve a model, but can simply discard models that lie beyond the threshold (Section 2.6 Filtering steric clashes). This caused a number of models to be dismissed due to clashes that never should have been modelled in the first place. The ideal would be a compromise between these two extremes that is efficient enough to run during the assembly algorithm, but also precise enough to eliminate all but minor surface clashes (Section 8.1.3 Refined clash detection). This will also benefit from a modelled representation of the components (Section 8.1.2 Atomic modelling). The structure is not the whole story, however. Some have suggested that functional annotation transfer is justified primarily for globular domains and requires caution when dealing with, e.g. transmembrane proteins. (Wong et al. 2010)

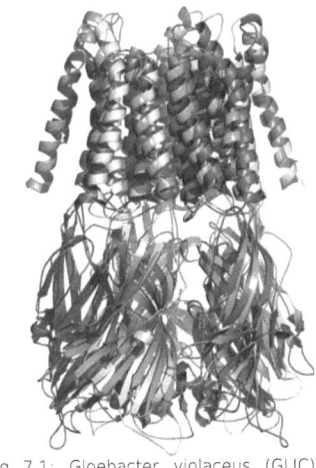

Fig 7.1: Gloebacter violaceus (GLIC) ion channel

Homopentameric ring of non-globular, multi-domain proteins. All five proteins modelled, 612 residues. C-alpha RMSD 5.33Å. Sequence identity 22%. Clash threshold 75%. Target ID: 3EHZ.

7.6 Interaction conservation

Only in the 3D Repertoire set did we incorporate homology models of the individual components. In general, a modelled complex is a scaffold that does not represent the protein sequences of interest. It contains protein chains from the homologous templates, possibly truncated to the length of the alignment to the template. Traditional monomeric homology modelling of the individual components is still required in order to be able to label the complex model as a model for the proteins under study (Section 8.1.2 Atomic modelling). We originally decided not to perform homology modelling because modelling tools generally do not take interfaces into account, with some exceptions (Fleishman et al. 2010). As a complex is assembled, each new superposition of shared components induces a new interface (Section 2.3.3 Merging complexes with shared components). If homology modelling tools do not consider this, then the interface, precisely the part that we are most interested in, may not be correctly retained, as the interface itself defines structural constraints that should be fulfilled.

This extent to which such modelling makes sense depends ultimately on how justified we are in assuming that a given interaction will be conserved. This depends upon our staying in the established boundaries of homology modelling, both monomeric (Fig 1.4: Applications of (monomeric) homology models at various levels of accuracy) and interface modelling (Fig 1.5: Transferability of functional annotation of protein-protein interactions). The results on our benchmark are consistent with previously published limits for interaction modelling (Section 1.4 Modelling interfaces), but we show that these limits also apply to multimeric complex modelling. This is a general guideline, only. Functional conservation ultimately has to be considered on a case-by-case basis.

7.7 Potential functional insights

One of the benchmark targets was resolved in the context of structural genomics initiatives, with limited functional annotation. However, the process of modelling its structure from homologous templates provided us with more detail, in a manner analogous to functional annotation of single proteins through homology detection. The homopentameric complex (PDB ID 1T0T) is from the Firmicute phylum and related to the family of chlorite dismutases. We identified three potential chlorite dismutase templates (Fig 7.2: Templates for unannotated benchmark target).

ID	Phylum	Identity	Length (protein)	Proteins
3DTZ	Archaea	19.20%	130	5
2VXH	Proteobacteria	22.30%	195	6
1VDH	Deinococcus-Thermus	54.10%	246	5

Fig 7.2: Templates for unannotated benchmark target

After making this observation, we discovered a newly released chlorite dismutase structure (ID 3NN1, 2010-07-28, (Kostan et al. n.d.)). While this proved to also be an inferior template (1.53Å over 915 atoms in 3NN1 versus 0.53Å over 1284 atoms in the original template 2VDH), the study also showed that a conserved arginine is present in all forms that have chlorite dismutase activity, whereas our target and template have a glutamine at this position (Fig 7.3: Chlorite dismutase-like family). This provided additional necessary evidence for what the structure seemed to suggest: the uncharacterised complex does not possess chlorite dismutase activity and seems most likely to have heme peroxidase activity. This demonstrates the contribution that our approach can make toward functional annotation of uncharacterised structural genomics complexes.

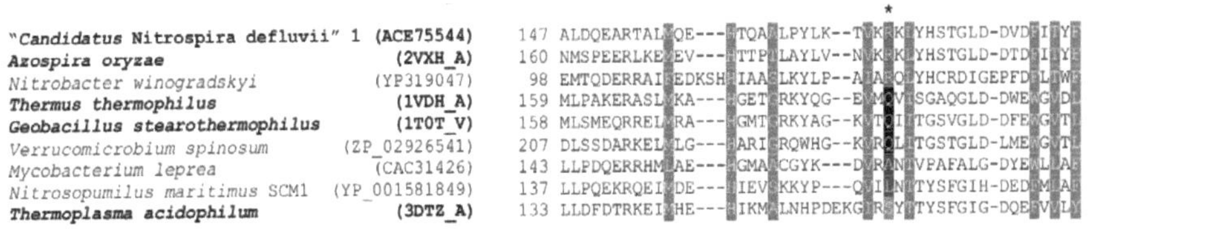

```
"Candidatus Nitrospira defluvii"  1  (ACE75544)     147 ALDQEARTAL QE---  TQA LPYLK--T K K YHSTGLD-DVD I Y
Azospira oryzae                      (2VXH_A)       160 NMSPEERLKE EV---  TTP LAYLV--N K K YHSTGLD-DTD I Y
Nitrobacter winogradskyi            (YP319047)       98 EMTQDERRAI EDKSH IAA LKYLP--A A Q YHCRDIGEPFD L W
Thermus thermophilus                 (1VDH_A)       159 MLPAKERASL KA---  GET RKYQG--E V SGAQGLD-DWE G D
Geobacillus stearothermophilus       (1TOT_V)       158 MLSMEQRREL RA---  GMT RKYAG--K T  TGSVGLD-DFE G T
Verrucomicrobium spinosum         (ZP_02926541)     207 DLSSDARKEL LG---  ARI RQWHG--K R L TGSTGLD-LME G T
Mycobacterium leprea                (CAC31426)     143 LLPDQERRHM AE---  GMA CGYK---D R N VPAFALG-DYE L A
Nitrosopumilus maritimus SCM1    (YP_001581849)     137 LLPQEKRQEI DE---  IEV KKYP---Q I N TYSFG1H-DED M A
Thermoplasma acidophilum             (3DTZ_A)       133 LLDFDTRKEI HE---  IKM LNHPDEKG R Y TYSFGIG-DQE V L
```

Fig 7.3: Chlorite dismutase-like family
*) R173 is conserved in all complexes with chlorite dismutase activity. Our target and template both have glutamine at this position. From (Kostan et al. n.d.)

7.8 Alternative conformations

Protein structures are not static. Proteins undergo conformational changes, which may result from, e.g. post-translational modifications (PTMs) or binding another protein, called *induced fit*. Such induced fit causes the structure of the protein at the interface to differ from the monomeric structure and is what makes docking of monomeric structures challenging (Lensink et al. 2007). Furthermore, a protein may have multiple bound conformations, if it binds to multiple partners. Our ability to model these different conformations depends on the existence of homologous interaction templates. It has been shown that multiple crystallisations of many proteins have led to an ensemble of structures with variations in the PDB (Burra et al. 2009; Hasegawa & Holm 2009; Kohn et al. 2010). This also covers induced fit of intrinsically disordered proteins (Tóth-Petróczy et al. 2008). Since we cluster the templates by structure, rather than by sequence, we implicitly take advantage of these alternative conformations in our models. This makes atomic modelling (Section 8.1.2 Atomic modelling) challenging, however. As mentioned (Section 2.3.3 Merging complexes with shared components), the interface in each dimeric template is real, but when we superimpose shared component structures, A' onto A'', the one structure A' has the one interface, e.g. A'--B, and its homologue A'' has the other interface, e.g. A''--C. The final model will contain three components, but there is no explicit interface A'--C (nor A''--B), as A' and C are simply juxtaposed. The extent to which these interfaces can be represented in detail, depends on the ability of the multi-template homology modelling to retain the details of both structures A' and A''. It should also be noted that what we have referred to as "assembly" here does not attempt to mimic the biological assembly process in vivo, which is a function of the evolution and the symmetry of the complex (Levy et al. 2008).

7.9 Nucleic acids

As we saw with the ribosomal small and large subunits (6.4 Predicted complex structures), neglecting nucleic acids can lead to gaps between components that make the model of little use. Nucleic acids play important structural and functional roles in complexes and are of interest in macromolecular modelling as much as proteins are. Theoretically, all of the algorithms developed here are applicable to protein-nucleotide interactions; none of them are protein-specific. Practically speaking, however, the tools and libraries upon which this project depends are very much protein-specific. An application to nucleic acids would require changes to: sequence databases, contact verification, sequence and structural alignments, interface template scoring, interface clustering, clash detection, model scoring, and benchmarking. For this reason, no nucleic acids were included in the benchmark complexes.

The relatively smaller number of protein-nucleotide templates in structure databases would also make this challenging. The extent to which protein-nucleotide interactions is conserved has also been shown to be more family-dependent and not as generalisable as for protein-protein interactions (Yu et al. 2004). Nevertheless, the current trends in nucleic acid research, particularly RNA structure, make this an inevitable development, including the tools that will be necessary to enable it.

8 Conclusion

As the volume of interaction data increases, the importance of complementing it with structural knowledge will grow. In order to study the multiple roles that proteins play, one must consider them in the different contexts in which they occur. We recognised the need to quantify these contexts by studying the structure of the different complexes in which each protein in a proteome participates. This requires homology modelling of structure that is context-dependent, on a scale that goes beyond current computational capacity, unless new approaches are developed. Methods for combining interaction structures manually have shown promise (Aloy et al. 2005). We automated and extended this to be capable of proteome-wide multimeric complex modelling. This allows us to consider every possible discrete complex that a set of proteins may form. This provides a first insight for all structurally plausible models, eliminating 99.99% of the impossible structures. We showed the ability of our approach to identify the correct complex on a set of benchmarks of known structure. Despite the context-dependent nature of the problem, we are able to produce the most realistic models early in the analysis. We have increased the sizes of complexes that combinatorial modelling can address (Section 1.5.3 Combinatorial docking) and have set new limits (Fig 6.2: Modelled complexes: 3D Repertoire)

From the *S. cerevisiae* proteome we were able to reconstruct the majority of complexes of known structure, particularly when the stoichiometry of the components was available. We were also able to provide plausible homology models for dozens of yeast complexes for which the complex structure remains unknown (Fig 6.2: Modelled complexes: 3D Repertoire, Fig 6.4: Modelled complexes: CYC2008). For each of these, we provide alternative complexes and model different isoforms, thus shedding light on the temporal aspect of complex interactions. We discovered that it is necessary to approach interaction networks hierarchically, modelling functional modules first, proceeding from the most stable toward the most dynamic, before considering attachment proteins or lower-confidence interactions. We also found that structural fit is essential for correct modelling and expect that our improvements in this area will increase the scope of our method, along with the rapid growth of interface structures (Tuncbag et al. 2008). Evaluating the individual models will be facilitated by a web interface that provides an overview of the sources of all of the data that contributed to building each model.

Overall, this work has provided a series of robust and unique tools for interrogating mechanism within protein interaction networks, and it together, with other methods to predict interaction mechanism (Puntervoll et al. 2003; Neduva et al. 2005; Lu et al. 2002; Diella et al. 2004; Linding et al. 2007) will serve to make the networks much more useful biochemically and ultimately lead both to a better understanding of what these increasingly complex networks mean, and provide useful points of intervention for studies of molecular function and disease.

8.1 Current work

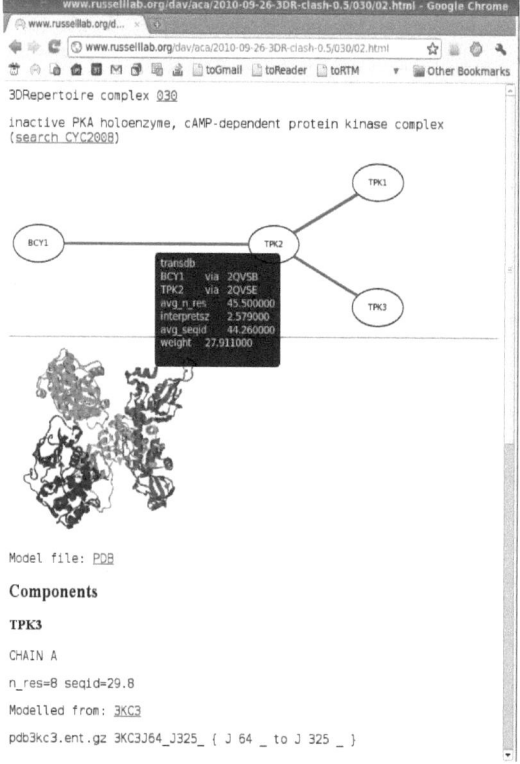

Fig 8.1: Web application
Showing a single model for cAMP-dependent protein kinase. The interaction network identifies which templates were used and links to more information on the interaction template. Below the graphic of the structure is a link to download the atomic coordinates, followed by the details of which templates were used to model which components. This is a preliminary web site and currently under active development.

8.1.1 Web services

Judging the reliability of a modelled complex requires information on the source species of the template interfaces, their biological function, the level of sequence similarity, the quality of the superpositions that put the model together, which interfaces were modelled and which were not, and a number of other factors (Fig 2.8: Scorable characteristics of complexes). In order to provide an overview for all of these data, we have been building a web application that presents statistics and annotations on the templates used in building a model. It also shows the interaction network of the model, i.e. the interfaces explicitly modelled and provides a view of the structure, identifying which structural chains correspond to which template sources (Fig 8.1: Web application). We will be expanding this with an interactive structural viewer and incorporating more information on the templates, to reduce the need to lookup data on other web sites. We will also provide programmatic access to our services, enabling them to be accessed from other programs. This platform is built on the software tools and libraries described in Section 8.2, Resources and tools developed. We will also catalogue all of our modelled yeast complexes here, along with other high-quality models, analogous to what ModBASE (Pieper et al. 2006; Pieper et al. 2008) has done for monomeric models.

8.1.2 Atomic modelling

As we discussed above (Section 7.6 Interaction conservation), the modelled complexes are scaffolds. In other words, these are templates for modelling a complex. They are not, however, a full homology model of the sequences themselves. Traditional homology modelling (Sali & Blundell 1993) must be performed at the sequence-by-sequence level, using the templates that are found for each complex. This is necessary to be able to retain as much atomic detail as possible, not just at one interface, but at every bound interface for each component in a complex.

While we have already done atomic modelling of individual components on 3D Repertoire complexes (Section 5.2 Structure of individual components), we do not yet do this in general. This will be automated by modelling each component sequence on the set of interfaces templates that it finds. Using multiple templates will allow the each component to better reflect the different interfaces (Section 7.8 Alternative conformations). A potentially more important reason, however, is that a modelled complex will allow us to do a more precise clash checking on the complexes that we assemble.

8.1.3 Refined clash detection

We found that our reduced representation was very efficient, but that the resolution was too low (Section 7.5 Clash detection) and that we could find no threshold that provided a balanced level of stringency. We also saw that doing refined clash detection after modelling only allows us to filter models, as it is then too late to improve them.

We will improve this by doing hierarchical clash detection. The current sphere-based representation of components will be extended, complementing this with a backbone-based clash detection. Both of these checks will happen during the traversal. Hierarchical means that the time-consuming backbone check will only be performed when the fast spherical check does not pass. This provides the best of both worlds: in the best case the method is no slower than the current method; in the worst case, we still get precise confidence that our model contains no steric clashes. We expect this to also increase the size of the complexes that we can model, as it will reduce the number of falsely identified clashes

8.1.4 Defining core complexes

We found that modelling the full-sized 3D Repertoire complexes proved to be too ambitious. We should have first focused our efforts on core complexes, before trying to model lower-confidence attachment proteins (Section 7.2 Complex composition). We saw this for example on the exosome, which contained 36 components, but for which we were not able to build the core hexameric ring structure. However, we had success in using seed models in order to build up to larger models (Section 6.3 Reconstruction of known complexes).

Here as well, we will take a hierarchical approach, beginning with the components annotated as belonging to the CORE set (Gavin et al. 2006) of 3D Repertoire. We will manually complement this set with the CYC2008 set to compensate for any missing core components. This set will be modelled using only high-confidence interactions, without docking. These models will then be extended with the remaining set of lower-confidence interactions from the larger 3D Repertoire set, in order to identify additional interactions. We will also make use of socio-affinity scores (Fig 5.1: Socio-affinities between complex components) to steer the search toward the most stable core components first (Section 8.1.6 Additional interaction data). This will ultimately result in a more directed search through the interaction network, improving not only the speed, but also the depth of our search. This two-stage modelling is only a first approximation for a more general problem: multimeric templates and N-stage modelling.

8.1.5 Hierarchical assembly

The motivation for a generalised hierarchical assembly is the observation that a template does not need to be limited to a dimeric structure (Alber et al. 2008). This reductionistic approach is what prevented us from being able to reconstruct some of the yeast complexes of known structure. Intuitively, if a tetrameric complex of known structure could serve as a template for e.g. four of the eight proteins that one would like to model, then the tetramer in its entirety should be used, rather then viewing the tetramer as as three (or more) dimeric templates and reassembling these anew. We saw that using multimeric seeds to build bigger models is a successful approach (Section 6.3 Reconstruction of known complexes). We also found (data not shown) that splitting query sequences into domains before modelling increased the number of target complexes that we could provide models for, at the expense of increasing the computational complexity.

Though intuitive, the problem is not trivial. Clearly, multimeric templates should not be reduced to their parts. The choice of the best template is ambiguous however, because it reintroduces the quality-versus-quantity trade-off. Is a larger template with a low identity preferable to smaller template with a high identity? Clearly, this depends on how much larger one template is and what level of identity the other has. This could continue down to the level of intra-chain domain-domain interfaces. What we are doing with multimeric complexes is analogous to what has been successfully performed on combinatorial fragment assembly of monomers (Simons et al. 1997; Simons et al. 1999; Simons et al. 1999; Haspel et al. 2003; Inbar et al. 2003). This will also speed up the assembly process, by not needing to reconstruct multimeric templates, and allow us to model larger, and more accurate complexes.

8.1.6 Additional interaction data

The current interaction database (Section 2.1 Structured interaction database) used interfaces from the PDB, from the asymmetric unit cell of each structure, but not from the biological units, which potentially include novel interfaces that could serve as interaction templates for complex modelling (Jefferson et al. 2006). First, we will adapt our sequence search to search biological quaternary structure of complexes, not simply the structures that are deposited in the PDB. Then, we will update our interface database to catalogue any new interfaces that are not in our current set.

We will also use the socio-affinity scores (Fig 5.1: Socio-affinities between complex components) from the 3D Repertoire complexes (when they are available), to prefer interactions between proteins with a stronger association. We will generalise this to permit up-weighting interactions by other experimental evidence. This will lead to a more directed search through the interaction network.

8.1.7 Novel interaction candidates

As we explicitly allow and look for ring topologies in complexes, there are candidates for potentially novel interactions for which no homologous template is known. In the 3D Repertoire set 1749 (24.5% of 7148) models had unidentified contacts. In the CYC2008 set 249 (26.0% of 956) models had unidentified contacts. Some of these may be explained through explicitly modelled interactions that did not happen until a later stage in the model building (Section 2.3.6 Detecting ring topologies). Still others may represent clashes that need to be filtered out. The remainder are the novel contacts that are not accounted for by known interactions. These are a consequence of the modelling and, in cases where the model is very reliable, provide leads for novel interfaces. The refined clash detection (Section 8.1.3 Refined clash detection), along with the atomic representation of components (Section 8.1.2 Atomic modelling) will make this list of candidates more definitive, so that we can validate these predicted interactions experimentally.

8.2 Resources and tools developed

Another important result of this project is the set of resources that came out of it and which continue to be used and improved. We have developed an interaction database that identifies conserved structural segments and homologous interfaces across the entire Protein Databank (2.1 Structured interaction database). We have defined a benchmark of known complexes that can be used to evaluate future complex modelling methods (Section 3.1 Defining a non-trivial benchmark). We have developed a library of software tools for structural bioinformatics that can be recombined to answer a number of other questions about interactions and structures (Section 8.2.1 Usage scenarios). We are developing a web site that will provide access to these tools as well as to our main complex modelling approach and to the models that we produce (Section 8.1.1 Web services).

8.2.1 Usage scenarios

The questions presented here can all be addressed by combining our existing tools:

- Given a set of proteins, which pairs have structural evidence for interacting?
- Given a set of proteins, which occur in different complex isoforms?
- Given a set of complex models, identify any potentially novel interactions, i.e. with no known homologous structures
- Given a complex structure, identify homologous interactions (interologues) in another structure.
- Given two (or more complexes), combine them using any homologous proteins that they may share.
- Given a complex structure and a set of proteins, which proteins can be modelled onto the complex and which are exclusive?
- Given two complexes, are they equivalent or is one a sub-complex of the other? If so, superimpose them.
- Given a structure, can it be extended by symmetry, e.g. to a ring topology.

8.2.2 Software libraries

We have developed a set of reusable structural bioinformatics libraries using modern role-based design (Fig 8.2: Structural bioinformatcs libraries). The software is implemented in Perl and uses the Perl Data Language (PDL, pdl.perl.org) for efficient transformation of large sets of atomic coordinates. Parallel processing and caching are integrated into every aspect of the libraries. The software is available under an open source licence from code.google.com/p/sbglib with documentation at russellab.org/aca/api.

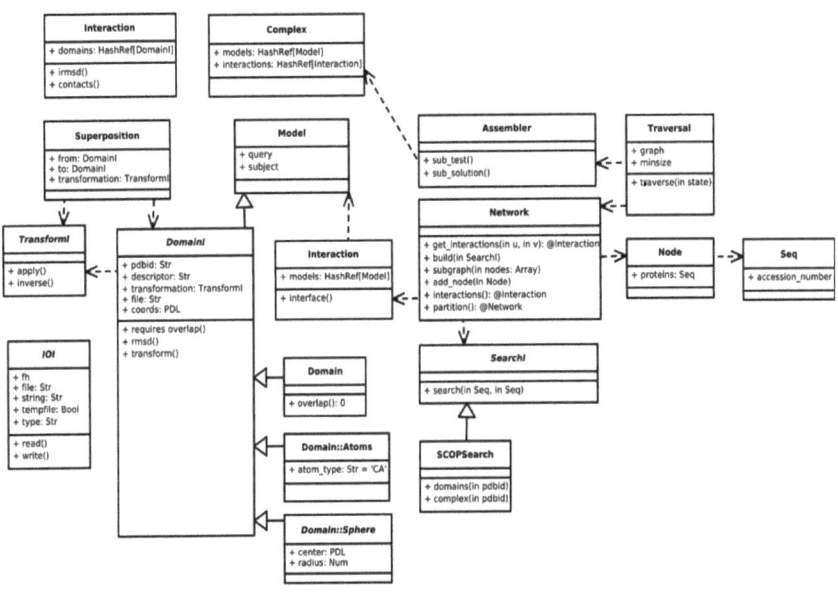

Fig 8.2: Structural bioinformatcs libraries

8.3 Contributions

Rob Russell and Patrick Aloy presented the original idea (Aloy et al. 2005), by extending their previous work on homology modelling of interactions (Aloy et al. 2003; Aloy & Russell 2003). Matthew Betts developed the database of interaction templates (unpublished, manuscript in preparation) and defined the preliminary set of benchmark complexes and provided strategies and advice on the complex modelling approach. Matthew also worked with Roberto Mosca on defining the yeast-specific interaction templates (Fig 5.5:Sources of yeast interface templates) (unpublished, manuscript in preparation). Roberto also provided the docking-based yeast interaction templates (Mosca et al. 2009). Matthieu Pichaud previously worked on another approach to complex modelling (Pichaud 2008) and provided general advice.

9 References

Agarwal, S. et al., 2010. Revisiting Date and Party Hubs: Novel Approaches to Role Assignment in Protein Interaction Networks. *PLoS Comput Biol*, 6(6), e1000817. Available at: http://dx.doi.org/10.1371/journal.pcbi.1000817 [Accessed June 21, 2010].

Alber, F., Kim, M.F. & Sali, A., 2005. Structural characterization of assemblies from overall shape and subcomplex compositions. *Structure*, 13(3), 435–445. Available at: http://dx.doi.org/10.1016/j.str.2005.01.013.

Alber, F. et al., 2007. The molecular architecture of the nuclear pore complex. *Nature*, 450(7170), 695–701. Available at: http://dx.doi.org/10.1038/nature06405.

Alber, F. et al., 2008. Integrating Diverse Data for Structure Determination of Macromolecular Assemblies. *Annu Rev Biochem*. Available at: http://dx.doi.org/10.1146/annurev.biochem.77.060407.135530.

Aloy, P. et al., 2004. Structure-based assembly of protein complexes in yeast. *Science*, 303(5666), 2026–2029. Available at: http://dx.doi.org/10.1126/science.1092645.

Aloy, P. et al., 2003. The relationship between sequence and interaction divergence in proteins. *J Mol Biol*, 332(5), 989–998. Available at: http://view.ncbi.nlm.nih.gov/pubmed/14499603.

Aloy, P., Pichaud, M. & Russell, R.B., 2005. Protein complexes: structure prediction challenges for the 21(st) century. *Curr Opin Struct Biol*, 15(1), 15–22. Available at: http://dx.doi.org/10.1016/j.sbi.2005.01.012.

Aloy, P. & Russell, R.B., 2003. InterPreTS: protein interaction prediction through tertiary structure. *Bioinformatics*, 19(1), 161–162. Available at: http://dx.doi.org/10.1093/bioinformatics/19.1.161.

Aloy, P. & Russell, R.B., 2002. Interrogating protein interaction networks through structural biology. *Proc Natl Acad Sci USA*, 99(9), 5896–5901. Available at: http://dx.doi.org/10.1073/pnas.092147999.

Aloy, P. & Russell, R.B., 2006. Structural systems biology: modelling protein interactions. *Nature Reviews Molecular Cell Biology*, 7(3), 188–197. Available at: http://dx.doi.org/10.1038/nrm1859.

Aloy, P. & Russell, R.B., 2004. Ten thousand interactions for the molecular biologist. *Nat Biotech*, 22(10), 1317–1321. Available at: http://dx.doi.org/10.1038/nbt1018.

Altschul, S.F. et al., 1997. Gapped BLAST and PSI-BLAST: a new generation of protein database search programs. *Nucleic Acids Res*, 25(17), 3389–3402. Available at: http://dx.doi.org/10.1093/nar/25.17.3389.

Amodeo, G.A., Rudolph, M.J. & Tong, L., 2007. Crystal structure of the heterotrimer core of Saccharomyces cerevisiae AMPK homologue SNF1. *Nature*, 449(7161), 492–495. Available at: http://dx.doi.org/10.1038/nature06127 [Accessed October 11, 2010].

Aranda, B. et al., 2010. The IntAct molecular interaction database in 2010. , 38(Database issue), D525-D531.

9 References

Baker, D. & Sali, A., 2001. Protein structure prediction and structural genomics. *Science (New York, N.Y.)*, 294(5540), 93-96. Available at: http://www.ncbi.nlm.nih.gov/pubmed/11588250 [Accessed September 6, 2010].

Bateman, A. et al., 2004. The Pfam protein families database. *Nucleic Acids Res*, 32 Database issue. Available at: http://view.ncbi.nlm.nih.gov/pubmed/14681378.

Baumeister, W., 2005. From proteomic inventory to architecture. *FEBS Lett*, 579(4), 933-937. Available at: http://dx.doi.org/10.1016/j.febslet.2004.10.102.

Berman, H. et al., 2007. The worldwide Protein Data Bank (wwPDB): ensuring a single, uniform archive of PDB data. *Nucleic Acids Research*, 35(Database issue), D301-D303. Available at: http://www.pubmedcentral.nih.gov/articlerender.fcgi?artid=1669775 [Accessed March 7, 2009].

Betts, M.J. & Sternberg, M.J., 1999. An analysis of conformational changes on protein-protein association: implications for predictive docking. *Protein Eng*, 12(4), 271-283. Available at: http://view.ncbi.nlm.nih.gov/pubmed/10325397.

Bi, X., 2007. Insights into COPII Coat Nucleation from the Structure of Sec23•Sar1 Complexed with the Active Fragment of Sec31. *Developmental Cell*, 13(5), 635-645. Available at: http://www.cell.com/developmental-cell/retrieve/pii/S153458070700384X [Accessed October 11, 2010].

Bi, X., Corpina, R.A. & Goldberg, J., 2002. Structure of the Sec23/24-Sar1 pre-budding complex of the COPII vesicle coat. *Nature*, 419(6904), 271-277. Available at: http://dx.doi.org/10.1038/nature01040 [Accessed October 11, 2010].

Bordner, A. & Gorin, A., 2008. Comprehensive inventory of protein complexes in the Protein Data Bank from consistent classification of interfaces. *BMC Bioinformatics*, 9(1). Available at: http://dx.doi.org/10.1186/1471-2105-9-234.

Bowman, G.D., O'Donnell, M. & Kuriyan, J., 2004. Structural analysis of a eukaryotic sliding DNA clamp-clamp loader complex. *Nature*, 429(6993), 724-730. Available at: http://dx.doi.org/10.1038/nature02585 [Accessed October 11, 2010].

Brohee, S. & van Helden, J., 2006. Evaluation of clustering algorithms for protein-protein interaction networks. *BMC Bioinformatics*, 7, 488+. Available at: http://dx.doi.org/10.1186/1471-2105-7-488.

Brooijmans, N., Sharp, K.A. & Kuntz, I.D., 2002. Stability of macromolecular complexes. *Proteins*, 48(4), 645-653. Available at: http://dx.doi.org/10.1002/prot.10139.

Burra, P.V. et al., 2009. Global distribution of conformational states derived from redundant models in the PDB points to non-uniqueness of the protein structure. *Proceedings of the National Academy of Sciences*, 106(26), 10505-10510. Available at: http://www.pnas.org/content/106/26/10505.abstract [Accessed July 7, 2009].

Ceol, A. et al., 2010. MINT, the molecular interaction database: 2009 update. , 38(Database issue), D532-D539.

Ceulemans, H. & Russell, R.B., 2004. Fast fitting of atomic structures to low-resolution electron density maps by surface overlap maximization. *J Mol Biol*, 338(4), 783-793. Available at: http://dx.doi.org/10.1016/j.jmb.2004.02.066.

Chandonia, J. & Brenner, S.E., 2006. The Impact of Structural Genomics: Expectations and Outcomes. *Science*, 311(5759), 347-351. Available at:

http://www.sciencemag.org/cgi/content/abstract/311/5759/347 [Accessed August 27, 2010].

Cheng, T.M., Blundell, T.L. & Fernandez-Recio, J., 2007. pyDock: Electrostatics and desolvation for effective scoring of rigid-body protein-protein docking. *Proteins: Structure, Function, and Bioinformatics*, 68(2), 503-515. Available at: http://onlinelibrary.wiley.com/doi/10.1002/prot.21419/abstract [Accessed October 9, 2010].

Cherry, J.M. et al., 1998. SGD: Saccharomyces Genome Database. *Nucleic Acids Research*, 26(1), 73-79.

Chothia, C., 1992. Proteins. One thousand families for the molecular biologist. *Nature*, 357(6379), 543-544. Available at: http://www.ncbi.nlm.nih.gov/pubmed/1608464 [Accessed August 30, 2010].

Collins, S.R. et al., 2007. Toward a Comprehensive Atlas of the Physical Interactome of Saccharomyces cerevisiae. *Mol Cell Proteomics*, 6(3), 439-450. Available at: http://www.mcponline.org/cgi/content/abstract/6/3/439 [Accessed January 18, 2010].

Cowieson, N.P., Kobe, B. & Martin, J.L., 2008. United we stand: combining structural methods. *Current Opinion in Structural Biology*, In Press, Corrected Proof. Available at: http://dx.doi.org/10.1016/j.sbi.2008.07.004.

Cyrklaff, M. et al., 2007. Whole Cell Cryo-Electron Tomography Reveals Distinct Disassembly Intermediates of Vaccinia Virus. *PLoS ONE*, 2(5), e420. Available at: http://dx.plos.org/10.1371/journal.pone.0000420 [Accessed October 15, 2010].

Darst, S., 1991. Three-dimensional structure of yeast RNA polymerase II at 16 Å resolution. *Cell*, 66(1), 121-128. Available at: http://www.cell.com/retrieve/pii/009286749190144N [Accessed October 11, 2010].

De Las Rivas, J. & de Luis, A., 2004. Interactome data and databases: different types of protein interaction. *Comparative and functional genomics*, 5(2), 173–178. Available at: http://dx.doi.org/10.1002/cfg.377.

Diella, F. et al., 2004. Phospho.ELM: a database of experimentally verified phosphorylation sites in eukaryotic proteins. *BMC Bioinformatics*, 5. Available at: http://dx.doi.org/10.1186/1471-2105-5-79.

Fath, S. et al., 2007. Structure and Organization of Coat Proteins in the COPII Cage. *Cell*, 129(7), 1325-1336. Available at: http://www.sciencedirect.com/science/article/B6WSN-4P2SJPT-G/2/ade33fc3856d2dcd0b430fcc64421166 [Accessed July 22, 2010].

Fleishman, S.J. et al., 2010. Rosetta in CAPRI rounds 13-19. *Proteins: Structure, Function, and Bioinformatics*, 9999(9999), NA. Available at: http://dx.doi.org/10.1002/prot.22784 [Accessed July 24, 2010].

Förster, F. et al., 2010. Toward an Integrated Structural Model of the 26S Proteasome. *Molecular & Cellular Proteomics*, 9(8), 1666 -1677. Available at: http://www.mcponline.org/content/9/8/1666.abstract [Accessed October 12, 2010].

Fukuhara, N. & Kawabata, T., 2008. HOMCOS: a server to predict interacting protein pairs and interacting sites by homology modeling of complex structures. *Nucleic acids research*, 36(Web Server issue). Available at: http://view.ncbi.nlm.nih.gov/pubmed/18442990.

Gagneur, J., David, L. & Steinmetz, L.M., 2006. Capturing cellular machines by systematic screens of protein complexes. *Trends in Microbiology*, 14(8), 336-339.

Gassel, M. et al., 2004. The Protein Kinase C Inhibitor Bisindolyl Maleimide 2 Binds with Reversed Orientations to Different Conformations of Protein Kinase A. *Journal of Biological Chemistry*, 279(22), 23679 -23690. Available at: http://www.jbc.org/content/279/22/23679.abstract [Accessed October 11, 2010].

Gavin, A.C. et al., 2002. Functional organization of the yeast proteome by systematic analysis of protein complexes. *Nature*, 415(6868), 141–147. Available at: http://dx.doi.org/10.1038/415141a.

Gavin, A.C. et al., 2006. Proteome survey reveals modularity of the yeast cell machinery. *Nature*, 440, 631–636. Available at: http://dx.doi.org/10.1038/nature04532.

Greene, L.H. et al., 2007. The CATH domain structure database: new protocols and classification levels give a more comprehensive resource for exploring evolution. *Nucleic Acids Res*, 35(Database issue). Available at: http://view.ncbi.nlm.nih.gov/pubmed/17135200.

Groll, M. et al., 1997. Structure of 20S proteasome from yeast at 2.4A resolution. *Nature*, 386(6624), 463-471. Available at: http://dx.doi.org/10.1038/386463a0 [Accessed October 10, 2010].

Guenther, S. et al., 2007. Docking without docking: ISEARCH-prediction of interactions using known interfaces. *Proteins*. Available at: http://dx.doi.org/10.1002/prot.21746.

Han, J.J. et al., 2004. Evidence for dynamically organized modularity in the yeast protein-protein interaction network. *Nature*, 430(6995), 88-93. Available at: http://dx.doi.org/10.1038/nature02555 [Accessed October 12, 2010].

Hasegawa, H. & Holm, L., 2009. Advances and pitfalls of protein structural alignment. *Current Opinion in Structural Biology*, In Press, Corrected Proof. Available at: http://www.sciencedirect.com/science/article/B6VS6-4WCWTNS-2/2/aca35507aef6eeafc142f8893db6762a [Accessed June 8, 2009].

Haspel, N. et al., 2003. Reducing the computational complexity of protein folding via fragment folding and assembly. *Protein Science*, 12(6), 1177-1187. Available at: http://dx.doi.org/10.1110/ps.0232903 [Accessed July 9, 2010].

Henrick, K. et al., 2003. EMDep: a web-based system for the deposition and validation of high-resolution electron microscopy macromolecular structural information. *Journal of Structural Biology*, 144(1-2), 228-237. Available at: http://www.sciencedirect.com/science/article/B6WM5-49SWBDF-4/2/9237505c240d1c6a7139634f4e388d3d [Accessed October 12, 2010].

Henrick, K. & Thornton, J.M., 1998. PQS: a protein quaternary structure file server. *Trends Biochem Sci*, 23(9), 358–361. Available at: http://view.ncbi.nlm.nih.gov/pubmed/9787643.

Ho, Y. et al., 2002. Systematic identification of protein complexes in Saccharomyces cerevisiae by mass spectrometry. *Nature*, 415(6868), 180-183. Available at: http://dx.doi.org/10.1038/415180a.

Huang, H., Jedynak, B.M. & Bader, J.S., 2007. Where Have All the Interactions Gone? Estimating the Coverage of Two-Hybrid Protein Interaction Maps. *PLoS Computational Biology*, 3(11), e214+. Available at: http://dx.doi.org/10.1371/journal.pcbi.0030214.

Humphrey, W., Dalke, A. & Schulten, K., 1996. VMD: Visual molecular dynamics. *Journal of Molecular Graphics*, 14(1), 33-38. Available at: http://www.sciencedirect.com/science/article/B6VNC-3VJRDJX-5/2/247595492d1cae80b34bed4d649bf0b2 [Accessed October 17, 2010].

Hunte, C. et al., 2000. Structure at 2.3 Å resolution of the cytochrome bc1 complex from the yeast Saccharomyces cerevisiae co-crystallized with an antibody Fv fragment. *Structure*, 8(6), 669-684. Available at: http://www.cell.com/structure/retrieve/pii/S0969212600001520 [Accessed October 11, 2010].

Inbar, Y. et al., 2005a. Combinatorial docking approach for structure prediction of large proteins and multi-molecular assemblies. *Phys Biol*, 2(4). Available at: http://view.ncbi.nlm.nih.gov/pubmed/16280621.

Inbar, Y. et al., 2005b. Prediction of multimolecular assemblies by multiple docking. *J Mol Biol*, 349(2), 435-447. Available at: http://dx.doi.org/10.1016/j.jmb.2005.03.039.

Inbar, Y. et al., 2003. Protein structure prediction via combinatorial assembly of sub-structural units. *Bioinformatics*, 19 Suppl 1. Available at: http://view.ncbi.nlm.nih.gov/pubmed/12855452.

Ito, T. et al., 2001. A comprehensive two-hybrid analysis to explore the yeast protein interactome. *Proc Natl Acad Sci U S A*, 98(8), 4569-4574. Available at: http://dx.doi.org/10.1073/pnas.061034498.

Jefferson, E.R., Walsh, T.P. & Barton, G.J., 2006. Biological units and their effect upon the properties and prediction of protein-protein interactions. *J Mol Biol*, 364(5), 1118-1129. Available at: http://dx.doi.org/10.1016/j.jmb.2006.09.042.

Jeffery, C.J., 1999. Moonlighting proteins. *Trends in biochemical sciences*, 24(1), 8-11. Available at: http://view.ncbi.nlm.nih.gov/pubmed/10087914.

Jensen, L.J. et al., 2008. STRING 8-a global view on proteins and their functional interactions in 630 organisms. *Nucl. Acids Res.*, gkn760+. Available at: http://dx.doi.org/10.1093/nar/gkn760.

Jung, S.H. et al., 2010. Protein complex prediction based on simultaneous protein interaction network. *Bioinformatics*, 26(3), 385-391. Available at: http://bioinformatics.oxfordjournals.org/cgi/content/abstract/26/3/385 [Accessed February 3, 2010].

Kabsch, W., 1976. A solution for the best rotation to relate two sets of vectors. *Acta Crystallographica Section A*, 32(5), 922-923. Available at: http://scripts.iucr.org/cgi-bin/paper?S0567739476001873 [Accessed October 5, 2010].

Karaca, E. et al., 2010. Building macromolecular assemblies by information-driven docking: introducing the HADDOCK multibody docking server. *Molecular & Cellular Proteomics: MCP*, 9(8), 1784-1794. Available at: http://www.ncbi.nlm.nih.gov/pubmed/20305088 [Accessed September 3, 2010].

Kim, P.M. et al., 2006. Relating three-dimensional structures to protein networks provides evolutionary insights. *Science*, 314(5807), 1938-1941. Available at: http://dx.doi.org/10.1126/science.1136174.

Kohn, J.E. et al., 2010. Evidence of Functional Protein Dynamics from X-Ray Crystallographic Ensembles. *PLoS Comput Biol*, 6(8), e1000911. Available at:

http://dx.doi.org/10.1371/journal.pcbi.1000911 [Accessed August 29, 2010].

Kostan, J. et al., Structural and functional characterisation of the chlorite dismutase from the nitrite-oxidizing bacterium "Candidatus Nitrospira defluvii": Identification of a catalytically important amino acid residue. *Journal of Structural Biology*, In Press, Corrected Proof. Available at: http://www.sciencedirect.com/science/article/B6WM5-50C71TV-1/2/2bff590dc66f96a9f7eaabf3a68925ad [Accessed October 8, 2010].

Kostelansky, M. et al., 2007. Molecular Architecture and Functional Model of the Complete Yeast ESCRT-I Heterotetramer. *Cell*, 129(3), 485-498. Available at: http://www.cell.com/retrieve/pii/S0092867407003790 [Accessed October 11, 2010].

Krissinel, E. & Henrick, K., 2007. Inference of macromolecular assemblies from crystalline state. *J Mol Biol*, 372(3), 774–797. Available at: http://dx.doi.org/10.1016/j.jmb.2007.05.022.

Krogan, N.J. et al., 2006. Global landscape of protein complexes in the yeast Saccharomyces cerevisiae. *Nature*. Available at: http://dx.doi.org/10.1038/nature04670.

Krumsiek, J., Friedel, C.C. & Zimmer, R., 2008. ProCope - Protein Complex Prediction and Evaluation. *Bioinformatics*, btn376+. Available at: http://dx.doi.org/10.1093/bioinformatics/btn376.

Kuehner, S. et al., 2009. Proteome Organization in a Genome-Reduced Bacterium. *Science*, 326(5957), 1235-1240. Available at: http://www.sciencemag.org/cgi/content/abstract/326/5957/1235 [Accessed June 30, 2010].

Kundrotas, P.J. & Alexov, E., 2006. Predicting 3D structures of transient protein-protein complexes by homology. *Biochim Biophys Acta*, 1764(9), 1498–1511. Available at: http://dx.doi.org/10.1016/j.bbapap.2006.08.002.

Lasker, K., Sali, A. & Wolfson, H.J., 2010. Determining macromolecular assembly structures by molecular docking and fitting into an electron density map. *Proteins: Structure, Function, and Bioinformatics*, n/a-n/a. Available at: http://onlinelibrary.wiley.com/doi/10.1002/prot.22845/abstract [Accessed August 29, 2010].

Lasker, K. et al., 2009. Inferential optimization for simultaneous fitting of multiple components into a cryoEM map of their assembly. *Journal of Molecular Biology*, In Press, Accepted Manuscript. Available at: http://www.sciencedirect.com/science/article/B6WK7-4VNH49X-8/2/de19b7cf7876644fd7f1f07a2978b743 [Accessed February 25, 2009].

Launay, G. & Simonson, T., 2008. Homology modelling of protein-protein complexes: a simple method and its possibilities and limitations. *BMC Bioinformatics*, 9(1). Available at: http://dx.doi.org/10.1186/1471-2105-9-427.

Lensink, M.F., M?ndez, R. & Wodak, S.J., 2007. Docking and scoring protein complexes: CAPRI 3rd Edition. *Proteins: Structure, Function, and Bioinformatics*, 69(4), 704-718. Available at: http://onlinelibrary.wiley.com/doi/10.1002/prot.21804/abstract [Accessed October 13, 2010].

Lensink, M.F. & Wodak, S.J., 2010. Blind predictions of protein interfaces by docking calculations in CAPRI. *Proteins: Structure, Function, and Bioinformatics*, 78(15), 3085-3095. Available at: http://onlinelibrary.wiley.com/doi/10.1002/prot.22850/abstract [Accessed October 12, 2010].

Levy, E.D. et al., 2008. Assembly reflects evolution of protein complexes. *Nature*. Available at:

http://dx.doi.org/10.1038/nature06942.

Lindert, S., Stewart, P.L. & Meiler, J., 2009. Hybrid approaches: applying computational methods in cryo-electron microscopy. *Current Opinion in Structural Biology*, 19(2), 218-225. Available at: http://www.sciencedirect.com/science/article/B6VS6-4VYKJPX-1/2/b4050c095b1a14450a891da96d9c974c [Accessed April 21, 2009].

Linding, R. et al., 2007. Systematic Discovery of In Vivo Phosphorylation Networks. *Cell*, 129(7), 1415-1426.

Liu, Q., Greimann, J.C. & Lima, C.D., 2006. Reconstitution, activities, and structure of the eukaryotic RNA exosome. *Cell*, 127(6), 1223-1237. Available at: http://dx.doi.org/10.1016/j.cell.2006.10.037.

Lu, L. et al., 2003. Multimeric threading-based prediction of protein-protein interactions on a genomic scale: application to the Saccharomyces cerevisiae proteome. *Genome Res*, 13(6A), 1146-1154. Available at: http://dx.doi.org/10.1101/gr.1145203.

Lu, L., Lu, H. & Skolnick, J., 2002. MULTIPROSPECTOR: an algorithm for the prediction of protein-protein interactions by multimeric threading. *Proteins*, 49(3), 350-364. Available at: http://dx.doi.org/10.1002/prot.10222.

Mewes, H.W. et al., 2004. MIPS: analysis and annotation of proteins from whole genomes. *Nucleic Acids Res*, 32 Database issue. Available at: http://view.ncbi.nlm.nih.gov/pubmed/14681354.

Mika, S. & Rost, B., 2006. Protein-Protein Interactions More Conserved within Species than across Species. *PLoS Computational Biology*, 2(7), e79+. Available at: http://dx.doi.org/10.1371/journal.pcbi.0020079.

Mosca, R. et al., 2009. Pushing Structural Information into the Yeast Interactome by High-Throughput Protein Docking Experiments. *PLoS Comput Biol*, 5(8), e1000490. Available at: http://dx.doi.org/10.1371/journal.pcbi.1000490 [Accessed September 8, 2009].

Mueller, M., Jenni, S. & Ban, N., 2007. Strategies for crystallization and structure determination of very large macromolecular assemblies. *Current Opinion in Structural Biology*, 17(5), 572-579. Available at: http://www.ncbi.nlm.nih.gov/pubmed/17964135 [Accessed August 31, 2010].

Murzin, A.G. et al., 1995. SCOP: a structural classification of proteins database for the investigation of sequences and structures. *J Mol Biol*, 247(4), 536-540. Available at: http://dx.doi.org/10.1006/jmbi.1995.0159.

Myer, V.E. & Young, R.A., 1998. RNA Polymerase II Holoenzymes and Subcomplexes. *Journal of Biological Chemistry*, 273(43), 27757-27760. Available at: http://www.jbc.org/content/273/43/27757.short [Accessed October 16, 2010].

Neduva, V. et al., 2005. Systematic Discovery of New Recognition Peptides Mediating Protein Interaction Networks. *PLoS Biol*, 3(12). Available at: http://dx.doi.org/10.1371/journal.pbio.0030405.

Neto, U.B. & Dougherty, E., 2004. Bolstered error estimation. *Pattern Recognition*, 37(6), 1267-1281. Available at: http://dx.doi.org/10.1016/j.patcog.2003.08.017.

Ning, K. et al., 2010. Examination of the relationship between essential genes in PPI network and hub proteins in reverse nearest neighbor topology. *BMC Bioinformatics*, 11(1), 505. Available at: http://www.biomedcentral.com/1471-2105/11/505 [Accessed October 14,

2010].

Nolen, B.J. & Pollard, T.D., 2007. Insights into the Influence of Nucleotides on Actin Family Proteins from Seven Structures of Arp2/3 Complex. *Molecular Cell*, 26(3), 449-457. Available at: http://www.cell.com/molecular-cell/retrieve/pii/S1097276507002535 [Accessed October 11, 2010].

Nussinov, R. & Wolfson, H.J., 1991. Efficient detection of three-dimensional structural motifs in biological macromolecules by computer vision techniques. *Proceedings of the National Academy of Sciences of the United States of America*, 88(23), 10495-10499. Available at: http://www.pnas.org/content/88/23/10495.abstract [Accessed March 6, 2009].

Orengo, C.A. et al., 2002. The CATH protein family database: a resource for structural and functional annotation of genomes. *Proteomics*, 2(1), 11-21. Available at: http://view.ncbi.nlm.nih.gov/pubmed/11788987.

Orengo, C.A. et al., 1997. CATH-a hierarchic classification of protein domain structures. *Structure*, 5(8), 1093-1108. Available at: http://view.ncbi.nlm.nih.gov/pubmed/9309224.

Oyedotun, K.S. & Lemire, B.D., 2004. The Quaternary Structure of the Saccharomyces cerevisiae Succinate Dehydrogenase. *Journal of Biological Chemistry*, 279(10), 9424-9431. Available at: http://www.jbc.org/content/279/10/9424.abstract [Accessed October 11, 2010].

Ozawa, Y. et al., 2010. Protein complex prediction via verifying and reconstructing the topology of domain-domain interactions. *BMC Bioinformatics*, 11(1), 350. Available at: http://www.biomedcentral.com/1471-2105/11/350 [Accessed July 24, 2010].

Park, S.Y. et al., 2004. In different organisms, the mode of interaction between two signaling proteins is not necessarily conserved. *Proc Natl Acad Sci U S A*, 101(32), 11646-11651. Available at: http://dx.doi.org/10.1073/pnas.0401038101.

Pichaud, M., 2008. *Protein Complexes Structure Prediction by Combination of Binary Interactions Derived by Homology*. Available at: http://archiv.ub.uni-heidelberg.de/volltextserver/volltexte/2008/8892/ [Accessed August 29, 2010].

Pieper, U. et al., 2006. MODBASE: a database of annotated comparative protein structure models and associated resources. *Nucleic Acids Res*, 34(Database issue). Available at: http://view.ncbi.nlm.nih.gov/pubmed/16381869.

Pieper, U. et al., 2008. MODBASE, a database of annotated comparative protein structure models and associated resources. *Nucleic acids research*. Available at: http://view.ncbi.nlm.nih.gov/pubmed/18948282.

Plewczynski, D. et al., 2010. VoteDock: Consensus docking method for prediction of protein-ligand interactions. *Journal of Computational Chemistry*, n/a-n/a. Available at: http://onlinelibrary.wiley.com/doi/10.1002/jcc.21642/abstract;jsessionid=9506B9EA8D5B BCC46907B19EC2B16101.d02t02 [Accessed October 12, 2010].

Puntervoll, P. et al., 2003. ELM server: A new resource for investigating short functional sites in modular eukaryotic proteins. *Nucleic Acids Res*, 31(13), 3625-3630. Available at: http://view.ncbi.nlm.nih.gov/pubmed/12824381.

Pu, S. et al., 2007. Identifying functional modules in the physical interactome of Saccharomyces cerevisiae. *Proteomics*, 7(6), 944-960. Available at: http://www.ncbi.nlm.nih.gov/pubmed/17370254 [Accessed September 5, 2010].

Pu, S. et al., 2009. Up-to-date catalogues of yeast protein complexes. *Nucleic Acids Research*, 37(3), 825-831. Available at: http://www.ncbi.nlm.nih.gov/pubmed/19095691 [Accessed September 5, 2010].

Rohl, C. et al., 2006. Cataloging the relationships between proteins: a review of interaction databases. *Mol Biotechnol*, 34(1), 69–93. Available at: http://view.ncbi.nlm.nih.gov/pubmed/16943573.

Rual, J. et al., 2005. Towards a proteome-scale map of the human protein-protein interaction network. *Nature*, 437(7062), 1173-1178. Available at: http://dx.doi.org/10.1038/nature04209 [Accessed October 15, 2010].

Russell, R.B. & Barton, G.J., 1992. Multiple protein sequence alignment from tertiary structure comparison: assignment of global and residue confidence levels. *Proteins*, 14(2), 309–323. Available at: http://dx.doi.org/10.1002/prot.340140216.

Sacher, M. et al., 2000. Identification and characterization of five new subunits of TRAPP. *European Journal of Cell Biology*, 79(2), 71-80. Available at: http://www.sciencedirect.com/science/article/B7GJ2-4DPMB00-6M/2/a2a71ccc53b30fff9c01c549ad78d838 [Accessed October 9, 2010].

Saeed, R. & Deane, C., 2007. An assessment of the uses of homologous interactions. *Bioinformatics*, btm576+. Available at: http://dx.doi.org/10.1093/bioinformatics/btm576.

Sali, A. & Blundell, T.L., 1993. Comparative protein modelling by satisfaction of spatial restraints. *J Mol Biol*, 234(3), 779-815. Available at: http://dx.doi.org/10.1006/jmbi.1993.1626.

Salwinski, L. et al., 2004. The Database of Interacting Proteins: 2004 update. *Nucleic Acids Res*, 32(Database issue). Available at: http://dx.doi.org/10.1093/nar/gkh086.

Shoemaker, B.A. & Panchenko, A.R., 2007a. Deciphering Protein-Protein Interactions. Part I. Experimental Techniques and Databases. *PLoS Computational Biology*, 3(3), e42+. Available at: http://dx.doi.org/10.1371/journal.pcbi.0030042.

Shoemaker, B.A. & Panchenko, A.R., 2007b. Deciphering Protein-Protein Interactions. Part II. Computational Methods to Predict Protein and Domain Interaction Partners. *PLoS Computational Biology*, 3(4), e43+. Available at: http://dx.doi.org/10.1371/journal.pcbi.0030043.

Simader, H. et al., 2006. Structural basis of yeast aminoacyl-tRNA synthetase complex formation revealed by crystal structures of two binary sub-complexes. *Nucleic Acids Research*, 34(14), 3968 -3979. Available at: http://nar.oxfordjournals.org/content/34/14/3968.abstract [Accessed October 11, 2010].

Simon, B. et al., 2010. An Efficient Protocol for NMR-Spectroscopy-Based Structure Determination of Protein Complexes in Solution. *Angewandte Chemie International Edition*, NA-NA. Available at: http://onlinelibrary.wiley.com/doi/10.1002/anie.200906147/abstract [Accessed August 24, 2010].

Simons, K.T. et al., 1997. Assembly of protein tertiary structures from fragments with similar local sequences using simulated annealing and bayesian scoring functions. *Journal of Molecular Biology*, 268(1), 209-225. Available at: http://www.sciencedirect.com/science/article/B6WK7-45VGF7T-N/2/7f41a915650d9e74c27a4f32afed75a4 [Accessed October 8, 2010].

9 References

Simons, K.T. et al., 1999. Improved recognition of native-like protein structures using a combination of sequence-dependent and sequence-independent features of proteins. *Proteins: Structure, Function, and Genetics*, 34(1), 82-95.

Simons, K.T. et al., 1999. Ab initio protein structure prediction of CASP III targets using ROSETTA. *Proteins*, Suppl 3, 171-176. Available at: http://www.ncbi.nlm.nih.gov/pubmed/10526365 [Accessed October 8, 2010].

Sinha, R., Kundrotas, P.J. & Vakser, I.A., 2010. Docking by structural similarity at protein-protein interfaces. *Proteins: Structure, Function, and Bioinformatics*, n/a-n/a. Available at: http://onlinelibrary.wiley.com/doi/10.1002/prot.22812/abstract [Accessed August 27, 2010].

Stein, A., Russell, R.B. & Aloy, P., 2005. 3did: interacting protein domains of known three-dimensional structure. *Nucleic Acids Res*, 33(Database issue). Available at: http://view.ncbi.nlm.nih.gov/pubmed/15608228.

Stein, A., Panjkovich, A. & Aloy, P., 2008. 3did Update: domain-domain and peptide-mediated interactions of known 3D structure. *Nucleic acids research*. Available at: http://view.ncbi.nlm.nih.gov/pubmed/18953040.

Stelzl, U. et al., 2005. A Human Protein-Protein Interaction Network: A Resource for Annotating the Proteome. *Cell*, 122(6), 957-968. Available at: http://www.sciencedirect.com/science/article/B6WSN-4H3YGBS-1/2/d39ee6e848fc3d640dcccc8f9ce59eaf [Accessed October 15, 2010].

Taverner, T. et al., 2008. Subunit Architecture of Intact Protein Complexes from Mass Spectrometry and Homology Modeling. *Acc. Chem. Res*. Available at: http://dx.doi.org/10.1021/ar700218q.

Teichmann, S.A., 2002. The constraints protein-protein interactions place on sequence divergence. *J Mol Biol*, 324(3), 399-407. Available at: http://view.ncbi.nlm.nih.gov/pubmed/12445777.

Teo, H. et al., 2004. ESCRT-II, an Endosome-Associated Complex Required for Protein SortingCrystal Structure and Interactions with ESCRT-III and Membranes. *Developmental Cell*, 7(4), 559-569. Available at: http://www.cell.com/developmental-cell/retrieve/pii/S1534580704003235 [Accessed October 11, 2010].

The UniProt Consortium, 2009. The Universal Protein Resource (UniProt) in 2010. *Nucleic Acids Research*, 38(Database), D142-D148. Available at: http://nar.oxfordjournals.org/content/38/suppl_1/D142 [Accessed October 4, 2010].

Thompson, J.D., Higgins, D.G. & Gibson, T.J., 1994. CLUSTAL W: improving the sensitivity of progressive multiple sequence alignment through sequence weighting, position-specific gap penalties and weight matrix choice. *Nucleic Acids Research*, 22(22), 4673-4680.

Topf, M. et al., 2005. Structural characterization of components of protein assemblies by comparative modeling and electron cryo-microscopy. *Journal of structural biology*, 149(2), 191-203. Available at: http://dx.doi.org/10.1016/j.jsb.2004.11.004.

Topf, M. et al., 2008. Protein Structure Fitting and Refinement Guided by Cryo-EM Density. *Structure*, 16(2), 295-307. Available at: http://dx.doi.org/10.1016/j.str.2007.11.016.

Tóth-Petróczy, Á. et al., 2008. Malleable Machines in Transcription Regulation: The Mediator Complex. *PLoS Comput Biol*, 4(12), e1000243. Available at: http://dx.doi.org/10.1371/journal.pcbi.1000243 [Accessed October 13, 2010].

Trabuco, L.G. et al., 2008. Flexible fitting of atomic structures into electron microscopy maps using molecular dynamics. *Structure (London, England: 1993)*, 16(5), 673-683. Available at: http://www.ncbi.nlm.nih.gov/pubmed/18462672 [Accessed May 28, 2010].

Tuncbag, N. et al., 2008. Architectures and Functional Coverage of Protein-Protein Interfaces. *Journal of molecular biology*, 381(3), 785-802.

Uetz, P. et al., 2000. A comprehensive analysis of protein-protein interactions in Saccharomyces cerevisiae. *Nature*, 403(6770), 623-627. Available at: http://dx.doi.org/10.1038/35001009.

Valencia, A. & Pazos, F., 2002. Computational methods for the prediction of protein interactions. *Curr Opin Struct Biol*, 12(3), 368-373. Available at: http://view.ncbi.nlm.nih.gov/pubmed/12127457.

Varma, S. & Simon, R., 2006. Bias in error estimation when using cross-validation for model selection. *BMC Bioinformatics*, 7(1). Available at: http://dx.doi.org/10.1186/1471-2105-7-91.

de Vries, S.J. et al., 2010. Strengths and weaknesses of data-driven docking in critical assessment of prediction of interactions. *Proteins: Structure, Function, and Bioinformatics*, n/a-n/a. Available at: http://onlinelibrary.wiley.com/doi/10.1002/prot.22814/abstract [Accessed August 24, 2010].

Walhout, A.J.M. et al., 2000. Protein Interaction Mapping in C. elegans Using Proteins Involved in Vulval Development. *Science*, 287(5450), 116-122. Available at: http://www.sciencemag.org/cgi/content/abstract/287/5450/116 [Accessed August 30, 2010].

White, C.L., Suto, R.K. & Luger, K., 2001. Structure of the yeast nucleosome core particle reveals fundamental changes in internucleosome interactions. *EMBO J*, 20(18), 5207-5218. Available at: http://dx.doi.org/10.1093/emboj/20.18.5207 [Accessed October 11, 2010].

Wong, P. et al., 2008. An evolutionary and structural characterization of mammalian protein complex organization. *BMC genomics*, 9(1). Available at: http://dx.doi.org/10.1186/1471-2164-9-629.

Wong, W., Maurer-Stroh, S. & Eisenhaber, F., 2010. More Than 1,001 Problems with Protein Domain Databases: Transmembrane Regions, Signal Peptides and the Issue of Sequence Homology. *PLoS Comput Biol*, 6(7), e1000867. Available at: http://dx.doi.org/10.1371/journal.pcbi.1000867 [Accessed August 6, 2010].

Wu, J. et al., 2007. PKA Type II{alpha} Holoenzyme Reveals a Combinatorial Strategy for Isoform Diversity. *Science*, 318(5848), 274-279. Available at: http://www.sciencemag.org/cgi/content/abstract/318/5848/274 [Accessed October 11, 2010].

Xu, Q. et al., 2006. ProtBuD: a database of biological unit structures of protein families and superfamilies. *Bioinformatics*, 22(23), 2876-2882. Available at: http://dx.doi.org/10.1093/bioinformatics/btl490.

Yu, H. et al., 2004. Annotation transfer between genomes: protein-protein interologs and protein-DNA regulogs. *Genome Res*, 14(6), 1107-1118. Available at: http://dx.doi.org/10.1101/gr.1774904.

Zacharias, M., 2010. Accounting for conformational changes during protein-protein docking. *Current Opinion in Structural Biology*, In Press, Corrected Proof. Available at: http://www.sciencedirect.com/science/article/B6VS6-4YH4P9Y-1/2/32b1d5905b6d45160d92a137f74149b2 [Accessed March 18, 2010].

Zweig, M.H. & Campbell, G., 1993. Receiver-operating characteristic (ROC) plots: a fundamental evaluation tool in clinical medicine. *Clin Chem*, 39(4), 561–577. Available at: http://view.ncbi.nlm.nih.gov/pubmed/8472349.

Die VDM Verlagsservicegesellschaft sucht für wissenschaftliche Verlage abgeschlossene und herausragende

Dissertationen, Habilitationen, Diplomarbeiten, Master Theses, Magisterarbeiten usw.

für die kostenlose Publikation als Fachbuch.

Sie verfügen über eine Arbeit, die hohen inhaltlichen und formalen Ansprüchen genügt, und haben Interesse an einer honorarvergüteten Publikation?

Dann senden Sie bitte erste Informationen über sich und Ihre Arbeit per Email an *info@vdm-vsg.de*.

Sie erhalten kurzfristig unser Feedback!

VDM Verlagsservicegesellschaft mbH
Dudweiler Landstr. 99   Telefon +49 681 3720 174
D - 66123 Saarbrücken   Fax   +49 681 3720 1749
www.vdm-vsg.de

Die VDM Verlagsservicegesellschaft mbH vertritt

Printed by Books on Demand GmbH, Norderstedt / Germany